Other Publications
Rainbows in Love.
Athena Press. London. 2010
ISBN 987 1 84748 728 5

THE SIMPLE COMPLEXITY OF NUMBER NINE

SAID HANY

authorHOUSE

AuthorHouse™ UK
1663 Liberty Drive
Bloomington, IN 47403 USA
www.authorhouse.co.uk
Phone: 0800.197.4150

© 2015 Said Hany. All rights reserved.

No part of this book may be reproduced, stored in a retrieval system, or transmitted by any means without the written permission of the author.

Published by AuthorHouse   09/01/2015

ISBN: 978-1-5049-8973-2 (sc)
ISBN: 978-1-5049-8972-5 (hc)
ISBN: 978-1-5049-8974-9 (e)

Print information available on the last page.

Any people depicted in stock imagery provided by Thinkstock are models, and such images are being used for illustrative purposes only.
Certain stock imagery © Thinkstock.

This book is printed on acid-free paper.

Because of the dynamic nature of the Internet, any web addresses or links contained in this book may have changed since publication and may no longer be valid. The views expressed in this work are solely those of the author and do not necessarily reflect the views of the publisher, and the publisher hereby disclaims any responsibility for them.

Dedication

To my grandchildren Aaliyah and Shara;
and to the memory of a unique man
and once in a lifetime friend,
Issa George Sawabini
(1947- 2015)

Contents

Dedication v
Acknowledgement ix
Introduction xi
History of Numbers 1
    The Number Sense 5
    The Basis of Our Number System 9
    History of Zero 11
    History of Pi (π) 15
    Pi (π) 15
    Characters of Sequence of Numbers 17
    Chronology of Numbers 17
    Types of Numbers 20
    Expanded List of Types of Numbers 25
    The Decimal Positional System 27
    Algorithm 27
    Prime Numbers 28
    Abbreviations Used in this Book: 29
    Arithmetic Operations 30
    Multiplication 31
    Numbers ending with a 9 42
    1 to 10 44
    Magic Number 45
    Multiplying Nines 46
    Quick multiplication by 9 46
    Addition 47
    369 49
    Subtraction 50
    Division 58
    Number Trees of 9 69
    Special Numbers (I) 71
    Reverse Multiplication 72
    Perfect Numbers 90

- The nine tables 91
- Central Nine! 91
- Special Numbers (II) 92
  - (1) - Fibonacci Numbers (Series) 92
  - The first 300 Fibonacci numbers 94
  - Every third cube number is a DR9! 129
- Nine Around the World 132
  - Cultures and Religions 133
  - Idioms and popular phrases 141
  - Literature 142
  - History 142
  - Mathematics 143
  - Geometry 150
  - Circle 151
  - Archaeology 153
  - Science 153
  - Sports 156
  - Languages 157
  - Music 157
  - Telecommunication 158
  - Miscellaneous 159

- References 163

Acknowledgement

Special thanks to my family for their support, encouragement and most of all patience; my wife Jean and my children Naysam, Nima and Ramaya.

Further thanks to *Miss Ramaya Hany* for her assistance and proofing of the mathematical mazes, and *Miss Evie McDermott* for her advice and suggestions with the historical and literal structure.

Introduction

"Numbers are not merely practical, but beautiful"…
The Pythagoreans

Number is the universal language of communication. All cultures around the world have, and share, numbers.

Numeration developed from quantities to numbers and from numbers to figures, where they are given names and are used according to universal principles that apply to all numbers.

The calculation and computing of numbers can only be accomplished with the invention of positional notation, using zero.

We find numbers everywhere around us and in every aspect of our life. Numbers are not just for the sciences and their branches, but also in arts and music, in beauty, harmony, balance, symmetry, and fantasies…

Number, with its universal language and internationally understood concepts, has the power to bridge all cultures.

The fascinating history of numbers and their development was always an attraction to me since early childhood.

I was born in Kumasi, Ghana to Lebanese parents. By the time I was three years old I was speaking three languages; Ashanti, Arabic and English. Often I wondered about the numbers in these three languages – baako, mmienu, mmiensa; wahad, ithnain, thalatha; one, two three; and kept repeating them to myself as I was falling asleep…

By the time I started high school I had a strong affinity to numbers, or what I came to realise later to be an obsession with numbers.

Out of that obsession, number nine would come back to me repeatedly.

Through my medical studies, I spent many lectures playing and messing with numbers and, nine was the number that surfaced every time…I could see nine, or the digital sum of 9, everywhere around me… In shops and restaurants, on cars, busses, trams and trains, on tickets and passes, in telephone and street numbers, in hours and minutes, and…

I kept notes of all my playful attempts with numbers, and took them with me around the world during my travels.

Since I retired, I reignited the urge to investigate number nine further, and to follow the history of numbers and enumeration.

Through my refreshed journey with numbers, I managed to discover patterns in some number sequences which have not been mentioned or reported before. Such numbers are the Fibonacci numbers, the Lucas numbers, cube numbers, prime numbers, and decimal expansion of the Golden Ratio and the Plastic Constant!

I often wonder whether numbers are an organised chaos or a chaotic organisation…

S. Hany,
August 2015.

History of Numbers

One, zero and infinity are the basis of the world of numbers.

Number 1 has been the first number used since man appeared on Earth and history began.

The first evidence of the existence of the number one, and that someone was using it to count, appeared about 20,000 years ago. It was just a series of lines cut into a bone, known as the Ishango Bone.

The Ishango Bone was found in the Democratic Republic of Congo of Africa in 1960. It is a fibula of a baboon, with a sharp piece of quartz affixed to one end, perhaps for engraving. On that bone, the lines cut into it are too uniform to be accidental. Indeed archaeologists believe, and agree, that the lines were marks to keep track of something.

In the 1970's during the excavations of Border Cave, a small piece of the fibula of a baboon, the *Lebombo bone*, was found marked with 29 clearly defined notches, and, at 37,000 years old, it ranks with the oldest mathematical objects known. The bone is dated approximately 35,000 BC and resembles the calendar sticks still in use by Bushmen clans in Namibia.

The Lebombo Bone is essentially a Baboon fibula that has tally marks on it. It is conjectured to have been used for tracking menstrual cycles, because it has 29 marks on it. It is older than the Ishango bone.

Numbers, and counting, did not truly come into being until the establishment of civilization and the rise of cities. Numbers, and counting, began about 4,000 BC in Sumer (Iraq now), one of the earliest civilizations.

The Egyptians transformed the number one from a unit of counting things to a unit of measuring, where in around 3,000 BC; the number one became used in Egypt as a unit of measurement of length.

The Egyptians were also the first civilization to invent different symbols for different numbers: The symbol for one was just a line; the symbol for ten was a rope, whilst he symbol for a hundred was a coil of rope. They also had numbers for one and ten thousand. The Egyptians were the first to dream up the number one million, where its symbol was a prisoner begging for forgiveness.

The earliest treatise on arithmetic that which we possess was written by an Egyptian priest, named Ahmes, between 1700 BC and 1100 BC, and even that is probably a copy of a much older work. It deals largely with the properties of fractions.

Greece made further contributions to the world of numbers and counting, much of it under the guidance of Pythagoras.

The Simple Complexity of Number Nine

It is to be noted that Pythagoras is actually a Phoenician who lived on the island of Samos in Greece. He studied in Egypt and Babylon, and upon returning to Greece, established a school of math, and later in Crotona in South Italy, introducing Greece to mathematical concepts already prevalent in Egypt. Pythagoras was the first man to come up with the idea of odd and even numbers. He taught that the entire universe is one vast system of mathematically correct combinations.

It is to be noted that the 'Yale Tablet' of Babylonia shows a great approximation of the square root of 2. Other tablets show that the Babylonians were well aware of the 'Pythagoras theorem'. It is to be noted that Pythagoras was transferred from Egypt to Babylon, as a prisoner of the Persian Emperor Cambyses in 525 BC. He stayed in Babylon for around 12 years where he came in contact with the mathematicians of Babylon and the magi of Zarathustra.

The next big advance in the world of numbers and mathematics came around 500 AD in India, and it would be the most revolutionary advance in numbers since the Sumerians invented math. The Indians invented an entirely new number: zero. It was they, who created a different symbol for every number from one to nine. These numbers are known today as Arabic numerals.

The Arabs had their figures from Hindustan, and never claimed the discovery for themselves.

The Indians have been using "Arabic" numbers then since about 500 BC.

Great were the merits of the Arabs in the advancement of mathematics; and especially in virtue of the fact that they preserved from oblivion the research results of both Greek and Hindu and handed them down to the Christian countries of the West.

Once zero was invented, it transformed counting, and mathematics, in a way that would change the world. The zero is still considered India's greatest contribution to the world; for the first time in human history, the concept of nothing had a number.

With the help of the very flexible Arabic numbers, Indian scientists worked out that the Earth spins on its axis, and that it moves around the sun, something that Copernicus would not discover for another thousand years.

The next big advance in numbers, the invention of fractions, came in 762 AD in what is now Baghdad, and what was then part of the Persian Empire. The Persians were Muslims, and it was their adherence to the Koran, and the teachings of Islam, that led to the invention of fractions.

The Koran taught that the possessions of the deceased had to be divided among their descendants. Unlike Christianity at the time, Islam, which was scarcely 100 years old at the time, divided belongings among women as well as men. In order to do this they required fractions. Prior to 762 AD they did not have a system of mathematics sophisticated enough to do a very proper job. Enter Arabic numbers.

Most of Europe switched from Roman to Arabic numerals in the Middle-Ages, partly due to Leonardo Fibonacci's work in which he praised the benefits of the Hindu-Arabic numeral system.

Islamic thinking was not far from the European minds at that time. The Muslims then ruled Sicily, North Africa, Malta, and Spain. When the Muslims were driven out of these places, they left behind their important mathematical and scientific legacy.

Most people forget, or do not know, that Islam was a more powerful culture and more scientifically advanced than the European civilizations in the centuries after the fall of the Roman Empire.

Baghdad was then a renowned learning center from Spain to India.

The Arabic numeral system is superior to the Roman one because it has a place system where the value of a number is determined by its position – positional notation.

The prototype of the number symbols we use today come from India. They are found in the Ashoka inscriptions from the 3rd century BC, the Nana Ghat a century later and in the Nasik Caves from the first century CE – all in shapes that considerably resemble our current symbols.

The Number Sense

The number sense is not the ability to count, but the ability to recognize changes and variations in a small collection. Some animal species are capable of this, indeed all mammals, and most birds, notice any change in the number of their young.

Children around 14-16 months of age will usually notice something that is missing from a group that he or she is familiar with. The same age child can usually reassemble objects again that have been separated into one group. However, the child's ability to perceive numerical differences in people or objects around him is very limited when the number goes beyond three or four.

The clearest idea of what counting and numbers mean may be gauged from the observation of children and of nations in the childhood of civilisation. When children count or add, they use either their fingers, or small sticks of wood, or pebbles, or similar things, which they adjoin singly to the things to be counted or otherwise ordinally associate with them. As we know from history, the Romans and Greeks employed their fingers when they counted or added. Moreover, even today we frequently meet with people to whom the use of the fingers is absolutely indispensable for computation.

The reason why the fingers are so universally employed as a means of numeration is, that everyone possesses a definite number of fingers, sufficiently large for purposes of computation and that they are always at hand.

We are born with the number sense, but we get to learn how to count.

The Indian numerals form the basis of the European number systems that are now widely used. These numerals were not transmitted directly from India to Europe, but came first to the Arabic/Islamic peoples and from them to Europe.

The eastern (Middle East) and western (North Africa and Spain) parts of the Arabic world both saw separate developments of Indian numerals with relatively little interaction between the two. Transmission to Europe came through this western Arabic route, coming into Europe first through Spain.

Al-Qifti, from Upper Egypt, in his *'Chronology of the Scholars'*, written around the end the 12th century, quotes much earlier sources:-

... a person from India presented himself before the Caliph al-Mansur in the year [776 AD] who was well versed in the siddhanta method of calculation related to the movement of the heavenly bodies, and having ways of calculating equations based on the half-chord [essentially the sine] calculated in half-degrees ...

This is all contained in a work from which he claimed to have taken the half-chord calculated for one minute. Al-Mansur ordered this book to be translated into Arabic, and a work to be written, based on the translation, to give the Arabs a solid base for calculating the movements of the planets...

The numbers were represented by letters but not in the dictionary order. The numbers from 1 to 9 were represented by letters, then the numbers 10, 20, 30, till 90 by the next nine letters (10 = y, 20 = k, 30 = l, 40 = m, and so on), then 100, 200, 300, etc., 900 by the next letters (100 = q, 200 = r, 300 = sh, 400 = ta, etc.). There are 28 Arabic letters and so one was left over which was used to represent 1000.

In fact, a closer look will show that between 969 and 1082 the biggest change in the numerals was the fact that the two and the three have been rotated through 90°. There is a reason for this change that came about due to the way that scribes wrote, for they wrote on a scroll which they wound from right to left across their bodies as they sat cross-legged. The scribes therefore, instead of writing from right to left (the standard way that Arabic is written), wrote in lines from top to bottom. The script was rotated when the scroll was read and the characters were then in the correct orientation.

Two Muslim mathematicians contributed greatly to the use of Indian numerals: al-Khwarizmi and al-Kindi of Persia, both of whom worked in the first half of the 9th century CE.

Greater were the merits of the Arabs in the advancement of mathematics; and especially in virtue of the fact that they preserved from oblivion the results of both Greek and Hindu research and handed them down to the Christian countries of the West. The Arabs expressly distinguished between the Archimedean approximate value and the two Hindu values: the square root of 10 and the ratio 62832:20000.

This distinction occurs also in Muhammad Ibn Musa Al-Khwarizmi, the same scholar who in the beginning of the ninth century brought the

principles of our present system of numerical notation from India and introduced it into the Islamic world. The Arabians, however, did not study the numerical quadrature of the circle only, but also the constructive. Ibn Alhaitham, who lived in Egypt about the year 1000, made an attempt of this kind. His treatise upon the squaring of the circle is preserved in a Vatican codex, which unfortunately has not yet been edited.

The first book written on the decimal system was written by Al-Khwarizmi which was translated into Latin under the title of '*Liber Algorismi de Numero Indorum*'. This book begins with the words '*dixit algorismi*' which means 'Al-Khwarizmi said'!

In 1437, al-Kashi defined decimal fractions, for which he proposed a simple notation, and established the rules of calculating with decimal numbers.

Using Arabic numbers Muslim mathematicians invented entirely new methods of mathematics. Beside just simple fractions they turned Arabic numbers into quadratic equations, and algebra, and these numeric breakthroughs enabled science, mathematics and astronomy to reach new levels in the Middle East.

The use of Arabic numerals in Europe is attributed to the Italian mathematician Fibonacci, where in 1202, he published a book called Liber Acci, which taught Arabic numerals and Algebra and, strongly advocated the use of Arabic numerals in society.

He was a merchant's son, born in Pisa in Italy, late in the twelfth century. In Pisa, he studied the work of Euclid and other Greek mathematicians. Whilst he was still a boy, he moved to the Muslim city of Bugia, in North Africa. There he examined leather and furs before they were shipped to Pisa.

Leonardo got an education in Arabic culture as he traveled around the Mediterranean to Greece, Constantinople, Egypt and Syria. He recognized that the Hindu-Arabic numerals, the numerals we use today, were superior to the Roman numerals he had grown up with in the West.

The Basis of Our Number System

The number system in use today is a place value decimal system meaning that it is not only the number, but also the placement of the number, is important. For example, number 351. This incorporates three numerals: 3, 5 and 1. Because we use a place value system, we know that the 3 does not stand just for 3, it means 300. The 5 stands for 50, and the 1, being in the ones place, is just 1. Rather than writing 300 + 50 + 1, our system allows us to write it simply as 351.

Our system is also a decimal one, because it is based upon increments of 10. We have 10 numerals in our system: 1, 2, 3, 4, 5, 6, 7, 8, 9 and 0. We count 1 through 9, and then move to the next level with 10. Then we go up in the ones place through 9 before moving to the next 10s place - 20. Each place value in the system is ten times the value of the one before it (Ones, tens, hundreds, thousands and so on).

The vital element in making this system work is the development of the concept of zero. Indeed the only group of people to develop the concept of zero, beside the Hindus, was the Mayans of Central America.

Logic was not the guiding light of the history of number systems. Numbers were invented and developed in response to the demands of accountants, priests, astronomers, astrologers, and mathematicians.

The majority of people throughout history failed to discover the rule of position, which was discovered in fact only four times in the history of the world. The rule of position is the principle of a numbering system in which a 9, let's say, has a different magnitude depending on whether it comes in first, second, third position in a numerical expression.

The first discovery of this essential tool of mathematics was made in Babylon, Iraq, in the second millennium BCE (Before Common Era). Chinese arithmeticians at around the start of the Common Era then rediscovered it. In the third to fifth centuries CE (Common Era), Mayan astronomers reinvented it, and in the fifth century CE, it was rediscovered for the last time, in India.

The Europeans were strongly attached to their archaic customs, and extremely reluctant to engage in new ideas. Many centuries passed before written arithmetic overtook the West.

Numbers should be distinguished from numerals which are the symbols representing numbers. And, a numeral system is a writing system which expresses numbers, i.e., a mathematical notation for representing numbers of a given set, using digits or other symbols in a consistent manner.

The equal '=' symbol was invented by Robert Recorde, a well-known mathematician, in 1557. He decided that two equal length parallel lines were as equal as anything available.

History of Zero

Zero is the nothing that does everything!

The number zero as we know it arrived in the West around 1200, delivered by the Italian mathematician Fibonacci (aka Leonardo of Pisa), who brought it, along with the rest of the Arabic numerals, back from his travels to North Africa. However, the history of zero, both as a concept and a number, stretches far deeper into history...

"There are at least two discoveries, or inventions, of zero," says Charles Seife, author of Zero: *The Biography of a Dangerous Idea*. "The one that we got the zero from came from the Fertile Crescent." It first came to be between 400 and 300 BC. in Babylon, Seife says, before developing in India, winding its way through northern Africa and, in Fibonacci's hands, crossing into Europe via Italy.

Initially, zero functioned as a mere placeholder - a way to tell 1 from 10 from 100. "That's not a full zero," Seife says. "A full zero is a number on its own; it's the average of -1 and 1."

The second appearance of zero occurred independently in the Mayan culture, in the first few centuries CE, which is the most striking example of the zero being devised wholly from scratch.

Other researchers pinpoint an even earlier emergence of a placeholder zero, a pair of angled wedges used by the Sumerians to denote an empty number column some 4,000 to 5,000 years ago.

From placeholder to the driver of calculus, zero has crossed the greatest minds and most diverse borders since it was born many centuries ago.

Today, zero is perhaps the most universally global symbol known. With zero, something can be made out of nothing.

Understanding and working with zero is the basis of our world today. Without zero, we would lack calculus, accounting, the ability to make arithmetic computations quickly, and, especially in today's connected world, computers.

The Sumerian system was handed down to the Akkadians around 2500 BC and then to the Babylonians in 2000 BC.

It was the Babylonians who first conceived of a mark to signify that a number was absent from a column; just as 0 in 1023 signifies that there are no hundreds in that number. Although zero's Babylonian ancestor was a good start, it would still be centuries before the symbol as we know it appeared.

Brahmagupta, around 650 AD, was the first to formalize arithmetic operations using zero. He used dots underneath numbers to indicate a zero. These dots were alternately referred to as 'sunya' (meaning empty), or 'kha' (meaning place). Brahmagupta wrote standard rules for reaching zero through addition and subtraction as well as the results of operations with zero. The only error in his rules was division by zero, which would have to wait for Isaac Newton and G.W. Leibniz to tackle.

However, it would still be a few centuries before zero reached Europe. First, the great Arabian voyagers would bring the texts of Brahmagupta and his colleagues back from India along with spices and other exotic items.

Zero reached Baghdad by 773 AD and was developed in the Middle East by Arabian mathematicians who would base their numbers on the Indian system.

In the ninth century, Mohammed ibn-Musa al-Khwarizmi was the first to work on equations that equaled zero, or algebra as it has come to be known.

Al-Khwarizmi's book *'hisab al-jabr wa al-muqabala'* was translated into Latin by Robert Chester in 1143. This book, the earliest introduction of Arabic numerals, not only provided the West with the concept of 'algebra' (derived from *'al-jabr'*), but also with the term *'algorithm'*, which is the misspelled name of Al-Khwarizmi.

A lunar impact crater located on the far side of the Moon is named after Al-Khwarizmi. It lies to the southeast of the crater Moiseev, and northeast of Saenger.

In the twelfth century, Latin translations of his work on the Indian numerals introduced the decimal positional number system to the Western world. His Compendious Book on Calculation by Completion and Balancing presented the first systematic solution of linear and quadratic equations in Arabic.

He also developed quick methods for multiplying and dividing numbers which was later called 'algorithm' (a corrupted derivation of his name). Al-Khwarizmi called zero 'sifr', from which cipher is derived. Zero was originally written as an oval in the 9th century, but was smaller than the other numbers. The Indians used pebbles for counting, and they noticed that when the last pebble was taken out, the mark left behind by the last pebble was a circle which, they adopted as the sign for 'nothingness'.

The first Indian name for zero was 'sunya' which is the Sanskrit name for emptiness. The Arabs derived their 'sifr' from 'sunya', which later became 'zephirum' in Latin and then 'zefiro', and eventually ended up in Europe as zero.

Zero reached Europe when the Arabs conquered Spain, and from there to England by the middle of the twelfth century through translations of Al-Khwarizmi's findings.

The first Western use of the digits, without the zero, was in the fifth century. Beothius, a Roman writer, explained in his geometry book how to operate the abacus using marked small cones instead of pebbles. The cones were marked with the symbols of Hindu-Arabic digits from one to nine and were called apices.

These 'apices' became the early representations of digits in Europe, and were given individual names: Igin for 1, Andras for 2, Ormis for 3, Arbas for 4, Quimas (or Quisnas) for 5, Caltis (or Calctis) for 6, Zenis (or Tenis) for 7, Temenisa for 8, and Celentis (or Scelentis) for 9. It is believed that some of these names originated from Arabic. On trying to decipher these names, I believe that Arbas was for four and Arba' was, and still is, for four in Arabic... And, Temenisa for eight, and Temenia was, and still is, for eight in Arabic...

The abacus had been the most prevalent tool to perform arithmetic operations at that time.

Fibonacci's developments were quickly noticed, and adopted, by Italian merchants and German bankers, and specially his use of zero.

Accountants knew their books were balanced when the positive and negative amounts of their assets and liabilities equaled zero. However, governments were still suspicious of Arabic numerals because of the ease in which it was possible to change one symbol into another. Though outlawed, merchants continued to use zero in encrypted messages, thus the derivation of the word cipher, meaning code, from the Arabic 'sifr'.

The next great mathematician to use zero was Rene Descartes, the founder of the Cartesian coordinate system. As anyone who has had to graph a triangle or a parabola knows, Descartes' origin is (0, 0). Although zero was now becoming more common, the developers of calculus, Newton and Leibniz, would make the final step in understanding zero.

Zero brings about the revolution in number thinking. To become the number we know now, it passed through three stages: symbol of notation, digit, and actual number.

> *"God made everything out of nothing, but the nothingness shows through"*
> P. Valery, French poet and philosopher.

History of Pi (ϖ)

(3.14159265358979323846...)

Pi is the ratio of the diameter of a circle to its circumference.

Since Pi is a non-ending, nonrepeating, infinite decimal, it has stood as a monument to futility and absolute uselessness. Various people have sat around calculating (without calculators, calculus, or even algebra) this number:

Ptolemy	(ca.150 AD)	3.1416
Tsu Ch'ung Chi	(430-501 AD)	355/113
Al Khwarizmi	(c. 800 AD)	3.1416
Al ' Kashi	(c. 1430)	14 places
Francois Viete	(1540-1603)	9 places
Adriaan van Roomen	(1561-1615)	17 places
Ludolph van Ceulen	(c. 1600)	35 places

Today you may download this number up to 50 million places on to your computer.

William Jones, from Wales in Britain, was the first to use π to denote pi in 1706.

Pi (ϖ)

=3.14159(26535897932384626)4338327950288419716939937510... Pi to first 50 digits.

Interestingly, there are many ancient codes inside Pi if you study the repeating numbers inside. In this example, there are 17 numbers inside the parenthesis. Notice the 26's and how the 7 is almost in the middle but instead it is the 9 that is dead center with 3 on its right. In addition, you can count exactly 9 digits on either side of the parenthesis to the center 9!

You can use a straight edge and compass to draw a circle and cross straight through the center.

Pi has a rich history in a wide range of fields, stretching from mathematics to the arts and pop culture. For example, in the world of mathematics he well-known Chinese mathematician Zu Chongzhi narrowed the value of Pi to an approximation of 355/113.

In 2006, Akira Haraguchi broke his own world record by memorizing pi to 100,000 decimal places…

In 1996, M. Keith wrote a short story 'Cadeic Cadenza', in which the word count matches the first 3,834 digits of pi…

Some uses of Pi:

- Describe the DNA double helix.
- Determining the distribution of primes.
- Analysing the ripples on water.
- Checking for accuracy - as there are now millions upon millions of known decimal places of Pi, by asking a super computer to compute this many figures its accuracy can be tested.
- In cryptography - the science of coding.
- Generate of a random number.

The most famous formula for calculating Pi is Machin's formula:

$$\pi/4 = 4 \arctan(1/5) - \arctan(1/239).$$

This formula, and similar ones, were used to push the accuracy of approximations to Pi to over 500 decimal places by the early 18th century (this was all hand calculation!).

Interestingly, there are no occurrences of the sequence 123456 in the first million digits of Pi!

Characters of Sequence of Numbers

A number sequence is a list of successive numbers that follow a certain pattern. The sequence can be finite or infinite.

1. Numbers progress in a single linear form, following one another.
2. The number that follows another number is obtained by adding 1.
3. Numbers get bigger and bigger.
4. They follow one another endlessly.
5. Although there is a first number, there is no last number…
6. They are in order, forming the archetype of order itself.

Chronology of Numbers

Discoveries, inventions and progress of numbers
(From 30,000 BC to 15th century AD – a century after the appearance of Zero in Europe)

BC

c. 30,000: Counting bones with numerical notches.

4th millennium: Appearance of calculations on clay stones in Mesopotamia and neighboring Middle East.

c. 3000: Writing, and first digits, invented in Sumer and Elam in the Middle East.

3rd millennium: Use of hieroglyphic numerals and additive decimal notation in Egypt.

c. 2700: Sumerian cuneiform digits in use.

c. 2000: Appearance of the decimal base system.

c. 1800: Positional notation appears in Babylon (Iraq).

c. 1700: The Rhind Papyrus in Egypt, in which the scribe Ahmes explores the measurement of the area of a circle.

c. 1300: Appearance of digits in China.

6th century: Pythagoras, Phoenician mathematical philosopher, founds the Pythagorean School in south Italy. The Pythagoreans distinguish between odd and even numbers.

5th century: Philosophers in Greece and Persia (Iran) use knotted ropes for calculation.

4th century: Aristotle develops first significant mathematical concept of infinity.

c. 300: Alexandria, in Egypt, becomes center of mathematics studies. Euclid in Alexandria writes the *Elements*, establishing mathematical method of postulate and proof.

3rd century:

- Appearance of first zero, in Babylon.
- Numerals partially resembling modern Indo-Arabic numerals appear in India.
- Archimedes writes a treatise on Pi (π).

2nd century: Use in China of positional numeration without the zero.

AD

c. 1st century: First use of negative numbers.

2nd century: Paper invented in China.

120: Chang Hing of China calculates Pi (π) as 3.1555…

4th - 5th century: Indian positional notation with zero.

5th-9th century: The Mayans use the place-system with zero.

6th century: Aryabhata of India refines the calculation of Pi to 3.1416.

8th century: Indian calculation method reaches Baghdad, beginning the golden age of Arabic mathematics.

9th century: Thabit ibn Qurrah, Arabic mathematician, describes amicable numbers.

c. 825: In Baghdad, Muhammad ibn Musa al-Khwarizmi, called the 'father of algebra', writes the *kitab al-jabrwa al-muqabalah* (Treatise on Restoration or Completion of Reduction or Balancing), on algebra.

10th century:
- The mathematician and poet Omar Khayyam, in Persia, develops a general number theory and publishes an influential treatise on algebra.
- Gobar numerals in use in North Africa and Spain which are the direct ancestors of the numerals used today in the West.

12th century: Arabic numerals, without zero, begin to appear in Europe, and become definitively in use there by the 14th century.

13th century:
- John of Halifax, English mathematician, writes *General Algorism*.
- In Peru, Inca *quipu*, knotted calculating cord, in use.
- In Mongolia, NasirEddin al-Tusi, writes on trigonometry and astronomy, and offers a proof of the parallel postulate.
  1202: Fibonacci writes the Liber Abaci (Book of the Abacus) on algebra.

14th century: The zero appears in Europe.

15th century:
- Development of the printing press.
- Indo-Arabic numerals acquire a definitive form in Europe.

- Negative numbers appear.
- Al-Kashi, in Samarkand, Uzbekistan, defines decimal fractions and writes *Miftah al-Hisab* (The Key to Arithmetic).

Types of Numbers

(1)- Main types

1-<u>Natural numbers</u>
The counting numbers (1, 2, 3...), are called natural numbers.

2-<u>Whole numbers</u>
They are the natural numbers and a zero. Not all whole numbers are natural numbers, but all natural numbers are whole numbers.

3-<u>Integers</u>
Positive and negative counting numbers, as well as zero. (...,-2,-1, 0, 1, 2...)

4-<u>Rational numbers</u>
Numbers that can be expressed as a fraction of an integer and a non-zero integer. All integers are rational, but the reverse is not true.

5-<u>Real numbers</u>
All numbers that can be expressed as the limit of a sequence of rational numbers. Every real number corresponds to a point on the number line. All rational numbers are real, but the reverse is not true.

6-<u>Irrational numbers</u>
A real number that is not rational is called irrational.

7-<u>Imaginary numbers</u>
Numbers that equal the product of a real number and the square root of minus 1.

8-<u>Complex numbers</u>
Includes real numbers, imaginary numbers, and sums and differences of real and imaginary numbers.

9-Hypercomplex numbers
Includes various number-system extensions: quaternions, octonions, tessarines, coquaternions, and biquaternions.

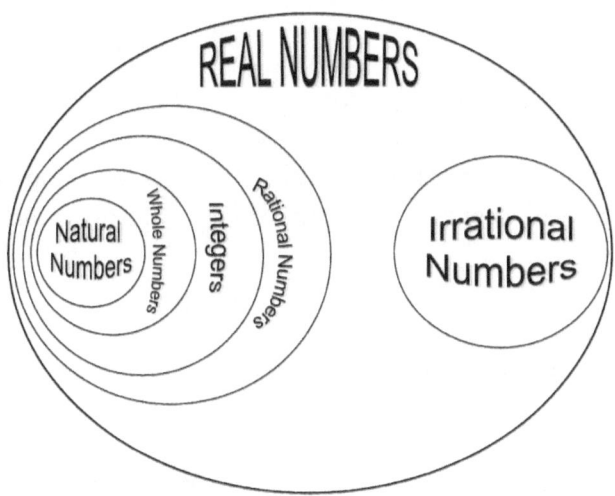

(2)-Number representations

1-Decimal
The standard Hindu–Arabic numeral system using base ten.

2-Binary
The base-two numeral system used by computers. See positional notation for information on other bases.

3-Roman numerals
The numeral system of ancient Rome, still occasionally used today.

4-FractionsA representation of a non-integer as a ratio of two integers. These include improper fractions as well as mixed numbers.

5-Scientific notation
A method for writing very small and very large numbers using powers of 10. When used in science, such a number also conveys the precision of measurement using significant figures.

(3)-Signed numbers

1-<u>Positive numbers</u>
Real numbers that are greater than zero.

2-<u>Negative numbers</u>
Real numbers that are less than zero.

Because zero, itself, has no sign, neither the positive numbers nor the negative numbers include zero. When zero is a possibility, the following terms are often used:

3-<u>Non-negative numbers</u>
Real numbers that are greater than or equal to zero. Thus, a non-negative number is either zero or positive.

4-<u>Non-positive numbers</u>
Real numbers that are less than or equal to zero. Thus, a non-positive number is either zero or negative.

(4)-Types of integers

An integer, from the Latin integer meaning 'whole', is a number that can be written without a fractional component.

1-<u>Even and odd numbers</u>
A number is even if it is a multiple of two, and is odd otherwise.

2-<u>Prime number</u>
A number with exactly two positive divisors.

3-<u>Composite number</u>
A number that can be factored into a product of smaller integers. Every integer greater than one is either prime or composite.

4-<u>Square number</u>
A number that can be written as the square of an integer.

There are many other famous integer sequences, such as the sequence of Fibonacci numbers, the sequence of factorials, the sequence of perfect numbers, and so forth.

Integer sequences which have their own name:

Abundant numbers	Juggler sequence
Baum–Sweet sequence	Kolakoski sequence
Bell numbers	Lucky numbers
Binomial coefficients	Lucas numbers
Carmichael numbers	Padovan numbers
Catalan numbers	Partition numbers
Composite numbers	Perfect numbers
Deficient numbers	Prime numbers
Euler numbers	Pseudoperfect numbers
Even and odd numbers	Pseudoprime numbers
Factorial numbers	Regular paper folding sequence
Fibonacci numbers	Rudin–Shapiro sequence
Figurate numbers	Semiperfect numbers
Golomb sequence	Semiprime numbers
Happy numbers	Superperfect numbers
Highly totient numbers	Thue-Morse sequence
Highly composite numbers	Ulam numbers
Home primes	Weird numbers
Hyperperfect numbers	

5-<u>Polygonal numbers</u>
These are numbers that can be represented as dots that are arranged in the shape of a regular polygon.

- Triangular numbers
- Square numbers
- Pentagonal numbers
- Hexagonal numbers
- Heptagonal numbers
- Octagonal numbers
- Nonagonal numbers
- Decagonal numbers
- Dodecagonal numbers (12 sides & 12 angles)

(5)-Algebraic numbers

1-Algebraic number
Any number that is the root of a non-zero polynomial with rational coefficients.

2-Transcendental number
Any real or complex number that is not algebraic. Examples include e and π.

3-Quadratic surd
It is an algebraic number that is the root of a quadratic equation. Such a number can be expressed as the sum of a rational number and the square root of a rational.

4-Constructible number
This is a number representing a length that can be constructed using a compass and straightedge. These are a subset of the algebraic numbers, and include the quadratic surds.

5-Algebraic integer
This integer is an algebraic number that is the root of a monic polynomial with integer coefficients.

(6)-Non-standard numbers

1-Transfinite numbers
Numbers that are greater than any natural number.

2-Ordinal numbers
Infinite numbers used to describe the order types of well-ordered sets. These include the cardinal numbers, which are used to describe the cardinalities of sets.

3-Infinitesimals
Nilpotent numbers. These are smaller than any positive real number, but are nonetheless greater than zero. These were used in the initial development of calculus, and are used in synthetic differential geometry.

4-Hyperreal numbers
The numbers used in non-standard analysis. These include infinite and infinitesimal numbers that possess certain properties of the real numbers.

5-Surreal numbers
A number system that includes the hyperreal numbers as well as the ordinals. The surreal numbers are the largest possible ordered field.

(7)-Computability and definability

1-Computable number
It is a real number whose digits can be computed using an algorithm.

2-Definable number
A real number that can be defined uniquely using a first-order formula with one free variable in the language of set theory.

Expanded List of Types of Numbers

Since numbers are infinite, then there will never ever be a complete list of numbers...

Integers	Digits
Numerals	Natural Numbers
Whole Numbers	Rational Numbers
Fractional Numbers	Irrational Numbers
Transcendental Numbers	Real Numbers
Abundant Numbers	Algebraic Numbers
Aliquot part	Almost perfect numbers
Alphametic Numbers	Amicable Numbers
Apocalypse Numbers	Arrangement Numbers
Automorphic Numbers	Beast Number
Binary Numbers	Cardinal Numbers
Catalan Numbers	Choice Numbers
Circular Primes	Complex Numbers
Composite Numbers	Congruent Numbers

Counting Numbers	Cubic Numbers
Cyclic Numbers	Decimal Numbers
Deficient Numbers	Digital Root
Egyptian fractions	Equable Triangles
Equivalent Numbers	Even Numbers
Factor Numbers	Factorial Numbers
Fermat Numbers	Fibonacci Numbers
Figurate Numbers	Golden Number (Golden Ratio)
Gnomon Numbers	Happy Numbers
Gyrating Numbers	Heronian Numbers
Hardy-Ramanujan Numbers	Infinite Numbers
Imaginary Numbers	Irrational Numbers
Integer Numbers	Least Deficient Numbers
Keith Numbers	Mersenne Numbers
Lucas Numbers	Multi-powered Numbers
Mono-digit Numbers	Narcissistic Numbers
Multiply-Perfect Numbers	Oblong Numbers
Natural numbers	Odd Numbers
Octahedral Numbers	Palindromic Numbers
Ordinal Numbers	Parasite numbers
Pan-digital Number	Pell Numbers
Pascal's Triangle Numbers	Perfect Numbers
Pentatope Numbers	Permutable Prime
Periodic Numbers	Polygonal numbers
Persistent Numbers	Prime Numbers
Powerful Numbers	Pronic Numbers
Product Perfect Numbers	Pyramidal Numbers
Pseudoprime Numbers	Quasiperfect Numbers
Pythagorean Numbers	Rational Numbers
Random Numbers	Rectangular Numbers
Real Numbers	Semi-perfect Numbers
Repunit Numbers	Sociable Numbers
Sequence Numbers	Square-full Numbers

Square Numbers	Superabundant Numbers
Squbes	Taxicab Numbers
Tag Numbers	Transcendental Numbers
Tetrahedral Numbers	Triangular Numbers
Trapezoidal Numbers	Triple Prime Numbers
Twin Prime Numbers	Unit fraction Numbers

The Decimal Positional System

The system that Fibonacci introduced into Europe used the Arabic symbols 1, 2, 3, 4, 5, 6, 7, 8, 9 with, most importantly, a symbol for zero 0. Using Roman numbers, 2001 could be written as MMI or it could be written as IMM - the order does not matter since the values of the letters are added to make the number in the original system. With the abbreviated system of IX meaning 9, then the order did matter but it seems this system was not often used in Roman times.

In the "new system", the order does matter always since 13 is quite a different number to 31. Also, since the position of each digit is important, then we may need a *zero* to get the digits into their correct places e.g. 2001 which has no tens and no hundreds - the Roman system would have omitted the values not used, so 'zero' was not needed.

This *decimal positional system*, as we call it, uses the ten symbols of Arabic origin, and the methods used by Indian Hindu mathematicians many years before they were imported into Europe. It has been commented that in India, the concept of nothing is important in its early religion and philosophy and so it was much more natural to have a symbol for it than for the Latin and Greek systems.

Algorithm

The Persian author Abu 'Abd Allah Mohammad ibn Musa al-Khwarizmi, usually abbreviated to Al-Khwarizmi, had written a book which included the rules of arithmetic for the decimal number system we now use, called Kitab al *jabr wa'l-muqabala* (Rules of restoring and equating) dating from about 825 AD. His name can be translated as Father of Abdullah, Mohammed, son of Moses, native of Khwarizm. He was an astronomer to the caliph at Baghdad.

Al-Khwarizm is the region south-east of the Aral Sea around the town now called Khiva (or Urgench) on the Amu Darya River. It was part of the famous Silk Route, a major trading pathway between the East and Europe. In the thirteenth century Al-Khwarizm was in Persia but today is in Uzbekistan, north of Iran, which gained its independence in 1991. Russia, then the USSR, issued a stamp in 1983 to commemorate al-Khwarizmi's 1200 year anniversary of his probable birth date. From the title of his book Kitab *al jabr wal-muqabala* we derive our modern word algebra. The Persian author's name is commemorated in the word algorithm. It has changed over the years from an original European pronunciation and latinisation of algorism.

The USA Library of Congress has a list of citations of Al-Khwarizmi and his works.

Prime Numbers

'Prime' originates from the Latin word 'primus' which means the first in importance.

A prime number can be divided, without a remainder, only by itself and by 1. For example, 7 can be divided only by 7 and by 1.

Some facts:

- The only even prime number is 2. All other even numbers can be divided by 2.
- If the sum of a number's digits is a multiple of 3, that number can be divided by 3.
- No prime number greater than 5 ends in a 5. Any number greater than 5 that ends in a 5 can be divided by 5.
- To prove whether a number is a prime number, first try dividing it by 2, and see if you get a whole number. If you do, then it can't be a prime number. If you don't get a whole number, next try

- dividing it by prime numbers: 3, 5, 7, 11 (9 is divisible by 3) and so on, always dividing by a prime number.
- Before 1951, before the dawn of computer wizardry, the largest known prime number was 44 digits long. This number was calculated by Lucas in 1876, and most probably will stand forever as the largest prime found by hand calculations.
- In 1951 Miller and Wheeler began using electronic computing and found several primes to reach a record of 79 digits.
- The largest known prime in the 21st century, $2^{43,112,609} - 1$, was found by electrical engineer Hans-Michael Elvenich on 6 Sept. 2008. It has 12,978,189 digits.
- But, 17,425,170-digit prime number was discovered in January 2013.
- Using previous data and graphs, we might guess that someone will discover:
  1-a 100,000,000 digit prime by late 2015, and
  2-a 1,000,000,000 digit prime by 2024.
- The sequence of primes starts with 2, 3, 5, 7, 11, 13, 17, 19, 23, and continues on from there forever to form the infinite set of prime numbers.
- Every integer greater than 1 is either prime or composite. For example, 7, 13, and 23 are prime (since each is only divisible by 1 and itself), while 8, 14, and 25 are all composite (since each is divisible by some number other than 1 and itself).

Abbreviations Used in this Book:

N or **No** = Number.
DS = Digital Sum.
DR = Digital Root.
O = Original number.
S = Sum.
L = Last number

Digital Sum (DS) and Digital Root (DR):

If we add up the digits of a number the result will be the digital sum and if we add up the digits of the digital sum, until there is only one number left, we have found what is called the digital root. In other words, the sum of the digits of a number is called its digital root.

327681 3+2+7+6+8+1 = 27 2+7 = 9
27 – Digital sum
9 – Digital root

Arithmetic Operations

The basic arithmetic operations are:

1- Addition
2- Subtraction,
3- Multiplication
4- Division

Addition (+)

Addition is the combination of two or more numbers into a single one, the sum of these numbers.

Subtraction (-)

Subtraction is the opposite of addition. Its purpose is to find the difference between two numbers, the minuend minus the subtrahend (minuend – subtrahend = difference)

If the minuend is larger than the subtrahend, the difference is positive; if the minuend is smaller than the subtrahend, the difference is negative; and if they are equal, then the difference is 0.

Multiplication (x or *)

Multiplication, like addition, combines two numbers into a single number, the 'product'. It is actually a repetition of addition. The two original numbers are called the 'multiplier' and the 'multiplicand', sometimes both simply called 'factors'.

Division (÷ or /)

Division is the inverse of multiplication. It finds the 'quotient' of two numbers, the 'dividend' divided by the 'divisor'. Any dividend divided by 0 is undefined. If the dividend is greater than the divisor, the quotient is greater than 1; otherwise it is less than 1. The quotient multiplied by the divisor always yields the dividend.

Dividend/Divisor = Quotient.

Divisor = Quotient/Dividend.

E.g. 28/4 = 7.   7x4 = 28.

MULTIPLICATION

9 x N = DR9

9 x 1 = 09   9 x 10 = 90
9 x 2 = 18   9 x 9 = 81
9 x 3 = 27   9 x 8 = 72
9 x 4 = 36   9 x 7 = 63
9 x 5 = 45   9 x 6 = 54
       135   252
(1+3+5=9) (2+5+2=9)

Note the *reversed similarity* between the left and right columns. 09 and 90, 18 and 81, 27 and 72, 36 and 63, and 45 and 54.

9 x 2 = 18.   1+8 = 9
9 x 3 = 27.   2+7 = 9
9 x 11 = 99.   9+9= 18.   1+8 = 9
9 x 23 = 207. 2+0+7 = 9
9 x 52 = 468. 4+6+8=18.   1+8 = 9
9 x 578329=5204961.   5+2+4+9+6+1= 27.   2+7 = 9
9 x 482729235601 = 4344563120409.
4+3+4+4+5+6+3+1+2+4+9 = 45.   4+5 = 9

170 x 9 = 1530 = 9
171 x 9 = 1539 = 18 = 9
172 x 9 = 1548 = 18 = 9
173 x 9 = 1557 = 18 = 9
174 x 9 = 1566 = 18 = 9
175 x 9 = 1575 = 18 = 9
176 x 9 = 1584 = 18 = 9
177 x 9 = 1593 = 18 = 9
178 x 9 = 1602 = 9
179 x 9 = 1611 = 9

DR9 x 9 = 9 or DR9

(4+5) x 9 = 81.   8+1 = 9
(2+7) x 9 = 81.   8+1 = 9
135 x 9 = 1215.   1+2+1+5 = 9
243 x 9 = 2187.   2+1+8+7 = 18. 1+8 = 9
428571 x 9 = 3857139
3+8+5+7+1+3+9 = 36.   3+6 = 9

DR9 x DR9 = DR9

63 x 36 = 2268.   2+2+6+8 = 18.   1+8 = 9.
45 x 27 = 1215.   1+2+1+5 = 9
18 x 63 = 1134.   1+1+3+4 = 9

Reverse multiplication of a number with a DR9 equals DR9

18 x 81 = 1458 = 18 = 9
27 x 72 = 1944 = 18 = 9
36 x 63 = 2268 = 18 = 9
45 x 54 = 2430 = 9

Quick multiplication by 999

As there are three nines in the number of nines, then place three zeros after the number to be multiplied. Then subtract the original number from the new one, which will result in the answer

Example 1: 63724 x 999.
Place the zeros, so 63724 become 63724000.

63724000 − 63724 = 63660276.
Hence, 63724 x 999 = 63660276!
Example 2: 35724 x 999
35724000 − 35724 = 35688276
Hence, 35724 x 999 = 35688276!

Similar numbers, which have a digital sum of 9, when multiplied by each other result in a digital root (DR) of 9, which is characterised by:

Outside numbers = 7
Middle numbers = 2
Total digital sum of all the resultant no. is = 9
Examples:
- 18 x 18 = 324

Outside numbers are 3 & 4.   3+4 = 7
Middle number of 324, is 2
DS: 3+2+4 = 9
- 27 x 27 = 729

Outside numbers are 7 & 9.   7+9 =16 =7
Middle number of 729 is 2
DS: 7+2+9 = 18   1+8 = 9
- 36 x 36 = 1296

Outside numbers are 1 & 6.   1+6 = 7
Middle numbers of 1296 are 29
2+9 = 11 1+1 = 2
DR: 1+2+9+6 = 18 1+8 = 9

(Nx9) + (N+9) = N9
Each and every 2-digit number that ends with a 9 is the sum of the multiple of the two digits plus the sum of the two digits. Thus, for instance:
39 = (3 x 9) + (3 + 9) = 27 + 12 = 39
59 = (5 x 9) + (5 + 9) = 45 + 14 = 59
69 = (6 x 9) + (6 + 9) = 54 + 15 = 69

89 = (8 x 9) + (8 + 9) = 72 + 17 = 89
99 = (9 x 9) + (9 + 9) = 81 + 18 = 99

Three Digit Reverse Multiplication = DR9
(N1N2N3 x N3N2N1 = DR9)
159 x 951 = 151209.   1+5+1+2+9 = 18 = 9
753 x 357 = 268821.   2+6+8+8+2+1 = 27 = 9
852 x 258 = 219816.   2+1+9+8+1+6 = 27 = 9

A repeated number when multiplied by 9 will result in a number that is equal to the single number multiplied by 9, with a 9, or 9s, in the middle.

Examples:

- 33 x 9 = 297 → 3 x 9 = 27. Place 9, once, in the middle of 2 and 7, resulting in 297.
- 444 x 9 = 3996 → 4 x 9 = 36. Place 9, twice, in the middle of 3 and 6, resulting in 3996.
- 5555 x 9 = 49995 → 4 x 9 = 36. Place 9 (3 times) in the middle of 4 and 5, resulting in 49995.

12345679 x 9 (or DR 9)
If 12345679 (all the numbers but without number eight) is multiplied by 1-9 multiples of 9, then the result must contain the numbers repeatedly.

Example:
12345679 x 9 = 111111111
12345679 x 18 = 222222222
12345679 x 27 = 333333333
12345679 x 36 = 444444444
12345679 x 45 = 555555555
12345679 x 54 = 666666666
12345679x 63 = 777777777
12345679 x 72 = 888888888
12345679x 81 = 999999999
12345679 x 90 = 1111111110

1 to 9

(1) - Sum starting with 1

```
123456789 x 9 =   1111111101=9
12345678  x 9 =    111111102=9
1234567   x 9 =     11111103=9
123456    x 9 =      1111104=9
12345     x 9 =       111105=9
1234      x 9 =        11106=9
123       x 9 =         1107=9
12        x 9 =          108=9
1         x 9 =            9=9
```

Sum ends with sequence 1 to 9.

(2) - Sum starting with sequence 1 to 9

```
(O)              (S)        (L) (DR)
123456789 x 9 = 1111111101=9
23456789 x 9 =   211111101=9
3456789 x 9 =     31111101=9
456789 x 9 =       4111101=9
56789 x 9 =         511101=9
6789 x 9 =           61101=9
789 x 9 =             7101=9
89 x 9 =               801=9
9 x 9 =                 81=9
```

Sum starts with same number as original (O).

Digital root (DR) of Sum (S) is never double numbers such as 18, 27, 36, etc.

- 2.1 - Sequence 1 to 5

```
123456789 x 9 = 1111111101 = 9
2345678 x 9 =      21111102 = 9
34567 x 9 =           311103 = 9
456 x 9 =               4104 = 9
5 x 9 =                   45 = 9
```

(3) - Sum starting with 2

23456789 x 9 = 211111101=9
2345678 x 9 = 21111102=9
234567 x 9 =   2111103=9
23456 x 9 =   211104=9
2345 x 9 =   21105=9
234 x 9 =   2106=9
23 x 9 =   207=9
2 x 9 =   18=9

Sum ends with sequence 1to 8

(4) - Sum starting with3

3456789 x 9 = 31111101 = 9
345678   x 9 = 3111102 = 9
34567   x 9 = 311103 = 9
3456   x 9 = 31104 = 9
345   x 9 = 3105 = 9
34   x 9 = 306 = 9
3   x 9 = 27 = 9

Sum ends with sequence 1 to 7

9 to 1

Sum starting with sequence 9 to 1

O (Original Number)   S (Sum)
98765432 x 9 =   8888888889
87654321 x 9 =   788888889
7654321 x 9 =   68888889
654321 x 9 =   5888889
54321 x 9 =   488889
4321 x 9 =   38889
321 x 9 =   2889
21 x 9 =   189
1 x 9 =   09

Digital sum is always equal to 9.
Examples: 8888888889 = 81 = 9
                    38889 = 36 = 9

The first number on the left in the sum (S) indicates the number of 8s to follow.

Example:
8888888889 – The first eight signifies that eight eights will follow.
788888889 – The 7 indicates seven eights will follow. Etc…

Sum starting with 9
Sum starts with 8.
Sum ends with sequence 9 to 2.

The number of 8's in the sum (S) is equal to the number of digits in the original number (O), which is also equal to the last number in that sum (L).

DS2 exhibits uniform sequence: 81 to 18 – from 8 to 1 (Left), & from 1 to 8 (Right).

(*)-888888889. First number on the left indicates the number of 8's to follow. (8)888888889.

O = Original   S = Sum   L = Last number

(O)		(S) (L) (DR2)
987654321	x 9 =	8888888889=81=9
98765432	x 9 =	888888888=72=9
9876543	x 9 =	88888887=63=9
987654	x 9 =	8888886=54=9
98765	x 9 =	888885=45=9
9876	x 9 =	88884=36=9
987	x 9 =	8883=27=9
98	x 9 =	882=18=9

Sum starting with 8
Sum starts with 7, and ends (L - Last number) with numbers from 9 to 2.

(O)   (S)   (L) (DS2)
87654321 x 9 = 788888889 = 81 = 9
8765432 x 9 =   78888888 = 72 = 9
876543   x 9 =   7888887 = 63 = 9
87654   x 9 =   788886 = 54 = 9
8765   x 9 =   78885 = 45 = 9
876   x 9 =   7884 = 36 = 9
87   x 9 =   783 = 27 = 9
8   x 9 =   72 = 18 = 9

Sum starting with 7
Sum starts with 6, and ends with sequence 9 to 3.
DS2 exhibits uniform sequence: 81 to 18 — from 8 to 1(Left), & from 1 to 8 (Right).

7654321 x 9 = 68888889 = 63 = 9
765432   x 9 =   6888888 = 54 = 9
76543   x 9 =   688887 = 45 = 9
7654   x 9 =   68886 = 36 = 9
765   x 9 =   6885 = 27 = 9
76   x 9 =   684 = 18 = 9
7   x 9 =   63 = 09 = 9

The first number on the left, of the result, indicates the number of eights to follow. Example: 68888889 = **63**. The six is followed by six eights.

6888888 = 54. The last eight in this number is in sequence with the number above, and the number below it.

688887 = 45. Four eights follow the six.

Sum starting with 6
Sum starts with 5, and ends with sequence 9 to 4.

The Simple Complexity of Number Nine

654321 x 9 = 5888889
65432 x 9 = 588888
6543   x 9 =   58887
654   x 9 =   5886
65   x 9 =   585
6   x 9 =   54

Sum starting with 5
Sum starts with 4, and ends with sequence 9 to 5.

54321 x 9 = 488889
5432   x 9 =   48888
543   x 9 =   4887
54   x 9 =   486
5   x 9 =   45

Sum starting with 4
Sum starts with 5, and ends with sequence 9 to 6.

4321   x 9 =   38889
432   x 9 =   3888
43   x 9 =   387
4   x 9 =   36

Sum starting with 3
Sum starts with 3, and ends with sequence 9 to 7.

321   x 9 = 2889 =27=9
32   x 9 =   288 =18=9
3   x 9 =   27 =09=9

Sum starting with 2
Sum starts with 2, and ends with sequence 9 to 8.

21 x 9 = 189
2 x 9 = 18

123456789

- 1x2x3x4x5x6x7x8x9 = 362880
  3+6+2+8+8 = 27 = 9

- 1x2x3x4x5x6 = 720 = 9
  1x2x3x4x5x6x7 = 5040 = 9
  1x2x3x4x5x6x7x8 = 40320 = 9

- 123456789 x 9 = 1,111,111,101 = 9!
  (9 x 123456789) - 123456789 =
  987654312

  9+8+7+6+5+4+3+1+2 = 45 = 9...

$$
\begin{array}{rl}
1 \times 9 = & 9 \\
2 \times 9 = & 18 \\
3 \times 9 = & 27 \\
4 \times 9 = & 36 \\
5 \times 9 = & 45 \\
6 \times 9 = & 54 \\
7 \times 9 = & 63 \\
8 \times 9 = & 72 \\
9 \times 9 = & 81 \\
\end{array}
$$

123456789 x 9 = **1**,111,111,101
 23456789 x 9 =   **2**11,111,101
  3456789 x 9 =   **3**1,111,101
   456789 x 9 =   **4**,111,101
    56789 x 9 =   **5**11,101
     6789 x 9 =   **6**1,101
      789 x 9 =   **7**,101
       89 x 9 =   **8**01
        9 x 9 =   81

The Simple Complexity of Number Nine

Note the similarity with the starting numbers…

123456789 x 9 = 1,111,111,101
12345678 x 9 =   111,111,102
1234567   x 9 =   11,111,103
123456   x 9 =   1,111,104
12345   x 9 =   111,105
1234   x 9 =   11,106
123   x 9 =   1,107
12   x 9 =   108
1   x 9 =   9

Note the opposing similarity with the end numbers.

987654321 x 9 = 8,888,888,889
98765432 x 9 =   888,888,888
9876543   x 9 =   88,888,887
987654   x 9 =   8,888,886
98765   x 9 =   888,885
9876   x 9 =   88,884
987   x 9 =   8,883
98   x 9 =   882
9   x 9 =   81

Note 1 to 9 on the left and 9 to 1 on the right.

987654321 x 9 = **8,888,888,889**
87654321 x 9 =   7**88,888,889**
7654321   x 9 =   6**8,888,889**
654321   x 9 =   5,**888,889**
54321   x 9 =   4**88,889**
4321   x 9 =   3**8,889**
321   x 9 =   2,**889**
21   x 9 =   1**89**
1   x 9 =   **9**

Again, note the pattern.

Similar numbers which have a digital sum of 9, when multiplied by each other result in:

- Outside numbers = 9
- Middle numbers = 2
- Total digital sum of all the resultant no. is = 9

Examples:

18 x 18 = 324 (3+4=7. 3+2+4=9)
27 x 27 = 729 (7+9=16=7. 7+2+9=18=9)
36 x 36 = 1296 (1+6=7. 2+9=11=2. 1+2+9+6=18=9)
45 x 45 = 2025 (2+5=7. 0+2=2. 2+0+2+5=9)

In the multiplications below, digits 1-9 were used once giving pandigital values:

9 x 57,624,831[1] = 1,037,246,958[2]
9 x 58,132,764 = 1,046,389,752
9 x 71,465,328 = 1,286,375,904
9 x 72,645,831 = 1,307,624,958
9 x 76,125,483 = 1,370,258,694
9 x 81,274,365 = 1,462,938,570

Sum of all numbers between 1 & 8 = 36 = 9

(1)- 5+7+6+4+8+3+1= 36, = 9
(2)- 1+0+3+7+2+4+6+9+5+8 = 36 = 9

Numbers ending with a 9

Every 2-digit number that ends with a 9 is equal to the multiple of the two digits plus the sum of the two digits.

Examples:

29
(2 x 9) + (2 + 9)
= 18 + 11 = 29

49
(4 x 9) + (4 + 9)
= 36 + 13 = 49!

59
(5 x 9) + (5 + 9)
= 45 + 14 = 59

79
(7 x 9) + (7 + 9)
= 63 + 16 = 79

89
(8 x 9) + (8 + 9)
= 72 + 17 = 89

Curious arrangements with 9's:

(0 x 9) + 1 =	1
(1 x 9) + 2 =	11
(12 x 9) + 3 =	111
(123 x 9) + 4 =	1 111
(1234 x 9) + 5 =	11 111
(12345 x9) + 6 =	111 111
(123456 x 9) + 7 =	1 111 111
(1234567 x 9) + 8 =	11 111 111
(12345678 x 9) + 9 =	111 111 111

The digital sum of the resulting number is equal to the added number.

E.g. in the second equation, the number added is **2**, and the resultant is 11 which has a digital sum of **2** (1+1), and so on…

1 to 10

1x2x3x4x5x6x7 = 5040. At the same time,
7x8x9x10 = 5040 (!!!)
5+0+4+0 = 9

The multiples of 9 are always composed of digits where the digital root is equal to 9; the product 123456789 x 9 gives 9 times the digit 1 in the answer (1111111101).

$$1 \times 9 = 9$$
$$2 \times 9 = 18$$
$$3 \times 9 = 27$$
$$4 \times 9 = 36$$
$$5 \times 9 = 45$$
$$6 \times 9 = 54$$
$$7 \times 9 = 63$$
$$8 \times 9 = 72$$
$$9 \times 9 = 81$$

$$1 \times 9 = 9$$
$$12 \times 9 = 10\mathbf{8}$$
$$123 \times 9 = 1 10 \mathbf{7}$$
$$1234 \times 9 = 11 10 \mathbf{6}$$
$$12345 \times 9 = 111 10 \mathbf{5}$$
$$123456 \times 9 = 1111 10 \mathbf{4}$$
$$1234567 \times 9 = 11111 10 \mathbf{3}$$
$$12345678 \times 9 = 111111 10 \mathbf{2}$$
$$123456789 \times 9 = 1111111 10 \mathbf{1}$$

The bold number indicates the number of times number one is repeated, to the left of the zero, in the answer...

Magic Number

To multiply 10,112,359,550,561,797,752,80
8,988,764,044,943,820,224,719

by 9; one just have to move the nine at the very end up to the front.

The results is:

9,101,123,595,505,617,977,528,089,887,640,449,438,222,471

This is the only number that does this!

Square of 9 or DR9 = 9 (DR9 x DR9 = 9)
9 x 9 = 81 = 9
27 x 27 = 729 = 18 = 9
36 x 36 = 1296 = 18 = 9
45 x 45 = 2025 = 9
63 x 63 = 3969 = 27 = 9
72 x 72 = 5184 = 18 = 9
135 x 135 = 18,225 = 18 = 9
612 x 612 = 374,544 = 27 = 9
4212 x 4212 = 17,740,944 = 36 = 9
3456 x 3456 = 11,943,936 = 36 = 9
91233 x 91233 = 8,323,460,289 = 45 = 9
720162 x 720162 = 518,633,306,244 – 45 – 9

Cube of 9 or DR9 = 9 (DR9 x DR9 x DR9 = 9)
9 x 9 x 9 = 729 = 18 = 9
36 x 36 x 36 = 46656 = 27 = 9
45 x 45 x 45 = 91125 = 18 = 9
135 x 135 x 135 = 2,460,375 = 27 = 9
1413 x 1413 x 1413 = 2,821,151,997 = 45 = 9

Multiplying Nines

Since it is easy to multiply by a power of 10, then this makes it easier to multiply any number by 9, 99 or any number of nines.

- 44 x 9:
  44 x 9 = 44 x (10-1) = 44 x 10 − 44
  = 440 − 44 = 440 − 40 − 4 = 396
  (440-40=400 400-4=396)
- 62 x 9:
  62 x 9 = 62 x (10-1) = 62 x 10 − 62
  = 620 − 62 = 620 − 60 − 2 = 558

Quick multiplication by 9

N_1N_2 x 9:
 N1N2 x 10 = X
 X − N1N2 = N1N2 x 9!

Examples:

1) - 9 x 23
    23 x 10 = 230
    230 - 23 = 207
    9 x 23 = 207!
2) - 9 x 55
    55 x 10 = 550
    550 − 55 = 495
    9 x 55 = 495!

0.999999999 = 1
- Proof 1:
  1/9 = 0.111…
  1/9 x 9 = 0.111 x 9
  9/9 = 0.999
  1 = 0.999!
- Proof 2:

$1/3 = 0.333...$
$2/3 = 0.666...$
$1/3 + 2/3 = 0.333 + 0.666$
$3/3 = 0.999$
$1 = 0.999...$
- Proof 3:
  If A is 0.999..., then 10A is 9.999...
  $10A - A = 9.999 - 0.999$
  $9A = 9$

Therefore A is equal to 1! Hence, 0.99 = 1!

Nine Ones:

111111111 x 111111111 =

1234567898 7654321

ADDITION

All Numbers
- 0123456789

$1+2+3+4+5+6+7+8+9 = 45$. $4 + 5 = 9$

<u>1 2 3 4</u>   5   <u>6 7 8 9</u>
  10   5   30

$1+0+5+3+0 = 45$. $4 + 5 = 9$

- 9 + (123456789 x 8) = 987654321!

123456789 = 45 = 9
<u>+12345678</u> = 36 = 9
135802467 = 36 = 9

N + 9 = DRN
1 + 9 = 10	1+0 = 1
3 + 9 = 12	1+2 = 3
25 + 9 = 34	
2 + 5 = 7	3+4 = 7
109 + 9 = 118	
109 1+0+9= 10	1+0 = 1
1+1+8 = 10 = 1	
346 + 9 = 22	2+2 = 4
3+4+6 = 13	1+3 = 4

N + Digits = DR9
342 + 3+4+2 = 351 =9
612 + 6+1+2 = 621 = 9
927 + 9+2+7 = 945 = 18 = 9

Any 2 numbers with a DS of 9 when added, the result is always equal to 9
18 + 27 = 45. 4+5 = 9
72 + 36 = 108. 1+ 0+8 = 9
171 + 162 = 333. 3+3+3 = 9
142857 + 758241 = 901098 = DS 27 = 9

The sum of the digits of the number added to 9 is always equal to the sum of the digits of the result.
Take any four-digit number, example:

9 + 17 = 26. 1 + 7 = 8. 2 + 6 = 8.

Intriguing sums:

Nine = nine + zero = eight + one = seven + two = six + three = five + four (the sums above contain all precisely 9 alphanumeric symbols)

```
  ↑  9 + 0  ↓
     8 + 1
     7 + 2
     6 + 3
     5 + 4
```

369

Nikola Tesla (10 July 1856 – 7 January 1943) was a Serbian American inventor, electrical engineer, mechanical engineer, physicist and futurist. He is best known for his contributions to the design of the modern alternating current (AC) electricity supply system.

Nikola Tesla believed in 369. He had said quite enigmatically, "If you only knew the magnificence of the 3, 6 and 9, then you would have a key to the universe."

- 1+2+4+5+7+8 = 27   2+7 = 9.
  At the same time, the missing numbers in the 6-sequence are 3, 6, and 9!

- 369 x 9
  3 x 6 x 9 = 162 = 9
  3 x 6   = 18 = 9
  3 x 9   = 27 = 9
  6 x 9   = 54 = 9

- 369 x 9 = 3321. 3+3+2+1 = 9
  3369 x 9 = 30321
  33369 x 9 = 300321
  3669 x 9 = 33021
  3666 x 9 = 330021
  3699 x 9 = 33291. 3+3+2+9+1 = 18 = 9
  36999 x 9 = 332991. 3+3+2+9+9+1 = 27 = 9

Two-digit Numbers Ending with 9

Each and every 2-digit number that ends with a 9 is the sum of the multiple of the two digits plus the sum of the two digits.

Thus, for instance:
- $39 = (3 \times 9) + (3 + 9)$
  $= 27 + 12 = 39$
- $79 = (7 \times 9) + (7 + 9)$
  $= 63 + 16 = 79$

Some three nines plus three nines equal 65934 (three nines)

```
   35964     53946    47952    41958
 + 29970    11988    17982    23976
   65934    65934    65934    65934…!!!
```

SUBTRACTION

Reverse Subtraction

Any random number (e.g. 35967930) when arranged with its integers in a descending order (i.e. 99765330) and subtracted from it the reverse number with rearranged integers in an ascending order (i.e. 03356799), the resulting subtraction (i.e. 96408531) individual integers when added (36) leads to a multiple of number 9.

Examples:

<u>2-Digit Numbers:</u>

= **9**
10-01 = 9
21-12 = 9
32-23 = 9
43-34 = 9
54-45 = 9
65-56 = 9
76-67 = 9

3-Digit Numbers:

=198. 1+9+8 = 18. 1+8 = **9**
210-012 = 198
321-123 = 198
432-234 = 198
543-345 = 198
654-456 = 198
765-567 = 198
876-678 = 198
987-789 = 198

4-Digit Numbers:

=3087. 3+0+8+7 = 18. 1+8=**9**
3210-0123 = 3087
4321-1234 = 3087
5432-2345 = 3087
6543-3456 = 3087
7654-4567 = 3087
8765-5678 = 3087
9876-6789 = 3087

5-Digit Numbers:

= 41976. 4+1+9+7+6 = 27. 2+7 =**9**
43210-01234 = 41976
54321-12345 = 41976
65432-23456 = 41976
76543-34567 = 41976
87654-45678 = 41976
98765-56789 = 41976

6-Digit Numbers:

= 530865. 5+3+0+8+6+5 = 27. 2+7 = **9**
543210 - 012345 = 530865
654321 - 123456 = 530865

765432 - 234567 =530865
876543 - 345678 =530865
987654 - 456789 =530865

7-Digit Numbers:

= 6419754. 6+4+1+9+7+5+4 = 36. 3+6 = **9**
6543210 - 0123456 = 6419754
7654321 - 1234567 = 6419754
8765432 - 2345678 = 6419754
9876543 - 3456789 = 6419754

8-Digit Numbers:

= 75308643 7+5+3+0+8+6+4+3 = 36. 3+6 = **9**
76543210 - 01234567 = 75308643   87654321 - 12345678 = 75308643
98765432 - 23456789= 75308643

9-Digit Numbers:

= 864197532. 8+6+4+1+9+7+5+3+2 = 45. 4+5 = **9**
876543210 - 012345678 = 864197532 987654321 -123456789 = 864197532

10-Digit Numbers:

9876543210 - 0123456789 = 9753086421 9+7+5+3+0+8+6+4+2+1 = 45. 4+5 = **9**
9988664432 -2344668899 = 7643995533 7+6+4+3+9+9+5+5+3+3 = 54. 5+4 = **9**

N - 9 = DR of No
14-9 = 5   1+4 = 5
212-9 = 203   2+1+2=5.   2+3=5

N – DSN = DR9
2356:
2+3+5+6 = 16

2356 − 16 = 2340 = 9
61420:
6+1+4+2 = 13
61420 − 13 = 61407 = 18 = 9

N − DRN = DR9
2356:
2+3+5+6 = 16 = 7
2356 − 7 = 2349 = 18 = 9
61420:
6+1+4+2 = 13 = 4
61420 − 4 = 61416 = 18 = 9
35:
3+5 = 8
35 − 8 = 27 = 9
88:
8+8 = 16 = 7
88 − 7 = 81 = 9

DR9 − DR9 = 9
18 − 9 = 9
1116 − 9 = 1107 = 9
1116 − 99 = 1017 = 9
1116 − 999 = 117 = 9
9999 − 1116 = 8883 = 27 = 9
6660 − 9 = 6651 = 18 = 9
6660 − 99 = 6561 = 18 = 9
13320 − 594 = 13914 = 18 = 9
9999 − 6660 = 3339 = 18 = 9
111111111 − 9 = 111111102 = 9

DS10 − DS10 (DR1 − DR1) = 9
37 − 28 = 9
82 − 55 = 27 = 9
271 − 181 = 90 = 9
2296 − 1729 = 567 = 18 = 9
2701 − 2296 = 405 = 9

$3025 - 2944 = 81 = 9$
$9128 - 8981 = 6147 = 18 = 9$

DRn - DRn = 9
- DR1 (DR1 – DR1 = 9)
    $28 - 19 = 9$ (2+8=10=1. 1+9=10=1)
    $37 - 28 = 9$
    $46 - 19 = 27 = 9$
    $55 - 28 = 27 = 9$
    $64 - 19 = 45 = 9$

- DR2 (DR2 – DR2 = 9)
    $74 - 38 = 36 = 9$ (7+4=11=2. 3+8=11=2)
    $101 - 2 = 99 = 18 = 9$
    $110 - 47 = 63 = 9$
    $110 - 65 = 45 = 9$
    $200 - 11 = 189 = 18 = 9$

- DR3 (DR3 – DR3 = 9)
    $30 - 12 = 18 = 9$ (3+0=3. 1+2=3)
    $111 - 48 = 63 = 9$
    $120 - 111 = 9$
    $210 - 57 = 153 = 9$
    $291 - 120 = 171 = 9$
- DR4 (DR4 – DR4 = 9)
    $31 - 13 = 18 = 9$ (3+1=4. 1+3=4)
    $40 - 22 = 18 = 9$
    $130 - 112 = 18 = 9$
    $2101 - 202 = 1899 = 27 = 9$
    $10210 - 2101 = 8109 = 18 = 9$
- DR 5 (DR5 – DR5 = 9)
    $41 - 23 = 18 = 9$ (4+1=5. 2+3=5)
    $311 - 50 = 261 = 9$
    $320 - 122 = 198 = 18 = 9$
    $4001 - 1310 = 2691 = 18 = 9$
    $6080 - 4010 = 2070 = 9$

- DR6 (DR6 − DR6 = 9)
  42 - 33 = 9 (4+2=6. 3+3=6)
  78 - 51 = 27 = 9
  411 - 222 = 189 = 18 = 9
  5010 - 3012 = 1998 = 27 = 9
  97143 - 6450 = 90693 = 27 = 9
- DR7 (DR7 − DR7 = 9)
  97 - 52 = 45 = 9 (9+7=16=7. 5+2=7)
  79 - 25 = 27 = 9
  322 - 52 = 270 = 9
  646 - 412 = 234 = 9
  5623 - 2707 = 2916 = 18 = 9
- DR8 (DR8 − DR8 = 9)
  62 - 44 = 18 = 9 (6+2=8. 4+4=8)
  98 - 62 = 36 = 9
  341 - 242 = 99 = 18 = 9
  5030 - 2321 = 2709 = 18 = 9
  41021 - 971 = 40050 = 9
- DR9 (DR9 − DR9 = 9)
  36 - 18 = 18 = 9 (3+6=9. 1+8=9)
  414 - 171 = 243 = 9
  630 - 405 = 225 = 9
  6102 - 3420 = 2682 = 18 = 9
  41301 - 2771 = 13590 = 18 = 9

DR9 - DR9 = 9 or DR9
90-63 = 27   2+7= 9 (9+0=9 & 6+3=9)
108-63 = 45   4+5 = 9
135-108 = 27   2+7 = 9
171-162 = 9
612-171 = 441 4+4+1 = 9
810-63 =747   7+7+4 =18 1+8 = 9
6428565 - 3571425 = 2857140
  6+4+2+8+5+6+5 = 36   3+6 = 9
  3+5+7+1+4+2+5 = 27   2+7 = 9
  2+8+7+1+4+0 = 27   2+7 = 9

DS10 - DS10 = DR9
82-55 = 27. 2+7 = 9 (8+2=10. 5+5=10)
271-181 = 90. 9+0 = 9
3025-1729 = 1296. 1+2+9+6 = 18. 1+8 = 9
6338-2855 = 3483. 3+4+3+8 = 18. 1+8 = 9
9128-2981 = 6147. 6+1+4+7 = 18. 1+8 = 9

N1N2 - N2N1= 9 or DR9
16 - 61 = -45. 4+5 = 9
19 - 91 = -72. 7+2 = 9
20 - 02 = 18. 1+8 = 9
21 - 12 = 9
26 - 62 = -36. 3+6 = 9
31 - 13 = 18. 1+8 = 9
32 - 23 = 9
41 - 14 = 27. 2+7 = 9
43 - 34 = 9
52 - 25 = 27. 2+7 = 9
54 - 45 = 9
63 - 36 = 27. 2+7 = 9
65 - 56 = 9
72 - 27 = 45. 4+5 = 9
75 - 57 = 18. 1+8 = 9
81 - 18 = 63. 6+3 = 9
82 - 28 = 54. 5+4 = 9
90 - 09 = 81. 8+1 = 9
91 - 19 = 72. 7+2 = 9

N1N2N3 - N2N3N1 = 9 or DR9
431 - 314 = 117. 1+1+7 = 9
628 - 286 = 342. 3+4+2 = 9
796 - 967 = -171. 1+7+1 = 9

N1N2N3 - N2N1N3 = 9 or DR9
623 - 263 = 360. 3+6 = 9
421 - 241 = 180. 1+8 = 9
546 - 456 = 90. 9+0 = 9

N1N2N3 - N1N3N2 = 9 or DR9
683 - 638 = 45. 4+5 = 9
564 - 546 = 18. 1+8 = 9
836 - 863 = -27. 2+7 = 9

N1N2N3 – N3N1N2 = 9 or DR9
431 - 143 = 288. 2+8+8=18. 1+8 = 9
735 - 573 = 162. 1+6+2 = 9
628 - 862 = 234. 2+3+4 = 9

N1N2N3 – N3N2N1 = 9 or DR9
625 - 526 = 99.   9+9 = 18 1+8 = 9
210 - 012 = 198   1+9+8 = 18 = 9
789 - 987 = -198. 1+9+8 = 18 = 9

N (2 or more digits) – DS = 9
20 - 2* = 18 = 9. 2+0 = 2*
59 - 5* = 63 = 9. 5+9 = 14. 1+4 = 5*
68 - 5* = 63 = 9. 6+8 = 14. 1+4 = 5*
77 - 5* = 72 = 9. 7+7 = 14. 1+4 = 5*
81 - 9* = 54 = 9. 8+1 = 9*
96 - 6* = 90 = 9. 9+6 = 15 = 6*

$(N_1N_2-N_1)-N_2 = 9$
46: (46-4)-6 =42-6=36=9
66: (66-6)-6 =60-6=54=9
85: (85-8)-5=77-5= 72=7+2=9
91: (91-9)-1=82-1=81=8+1=9

N minus Digital Sum of N is always equal to 9

Example: 6218

6 + 2 + 1 + 8 = 17
6218 – 17 = 6201 (6+2+1=9)

Example: 3684

3 + 6 + 8 + 4 = 21
3684 − 21 = 3663 (3+6+6+3=18=9)

Two nines minus two nines equal two nines equal DR9

8172 − 2718 = 5454 = 9 + 9 = 18 = 9
(8+1=9 7+2=9)

Three nines minus three nines equal three nines

```
 53946     47952     41958     35964
-11988    -17982    -23976    -29970
 41958*1   29970²    17982³    5994⁴
```
(3 nines) (3 nines) (3 nines) (3 nines)
(*) - 4+5, 1+8 & 9 = 3 nines... Similar other numbers...
(1)- 41958 = 666 x 63
(2)- 29970 = 666 x 45
(3)- 17982 = 666 x 27
(4)- 5994 = 666 x 9

DIVISION

Numbers divisible by 9

Numbers are divisible by 9 if the sum of all the digits is evenly divisible by 9. For example, the digital root of 4968 is 27 (4+9+6+8=27=9), which is divisible by 9 (27/9=3) so 4968 is evenly divisible by 9 (4968/9=552).

4968 / 9 = 552
4986 / 9 = 554
4869 / 9 = 541
4896 / 9 = 544
4698 / 9 = 522
4689 / 9 = 521

9864 / 9 = 1096
9846 / 9 = 1094
9684 / 9 = 1076
9648 / 9 = 1072
6984 / 9 = 776
6948 / 9 = 772
6498 / 9 = 722
6489 / 9 = 721
8964 / 9 = 996
8946 / 9 = 994
8694 / 9 = 966
8649 / 9 = 961

Divisible by 9

Is the number divisible by 9? This is the same as asking whether the number is in the '9 times table'.

Find the digital sum, then the digital root of that number. If the digital root come up to be 9, then the original number is divisible by 9.

Example: 817236

Digital sum: 8+1+7+2+3+6 = 27
Digital root: 2+7 = 9

As the digital root is 9, then the original number must be divisible by 9.

Indivisible by 9

Any number that is not divisible by nine, then the numbers after the decimal point will be repeated.

Example:

8 / 9 = 0.8888888888...
23 / 9 = 2.5555555555...

698 / 9 = 775.6666666666...
56768231 / 9 = 6307581.2222222222...

9 or DR9 divided by 5 equals 9 or DR9
9 / 5 = 1.8 (1+8=9)
18 / 5 = 3.6 (3+6=9)
27 / 5 = 5.4 (5+4=9)
36 / 5 = 7.2 (7+2=9)
135 / 5 = 27 (2+7=9)
171 / 5 = 34.2 (3+4+2=9)
432 / 5 = 86.4 (8+6+4=18=9)
621 / 5 = 124.2 (1+2+4+2=9)
711 / 5 = 142.2 (1+4+2+2=9)
801 / 5 = 160.2 (1+6+2=9)
900 / 5 = 180 (1+8=9)

Any DR9 is divisible by 9
18 / 9 = 2
45 / 9 = 5
108 / 9 = 12
135 / 9 = 17
144 / 9 = 16
198 / 9 = 22
207 / 9 = 23
261 / 9 = 29
342 / 9 = 38
405 / 9 = 45
1503 / 9 = 167
5121 / 9 = 569
5112 / 9 = 568
6003 / 9 = 667
6111 / 9 = 679
7101 / 9 = 789
41022 / 9 = 4558

DR9 divided by two = DR9

9 / 2 = 4.5   4+5 = 9
252 / 2 = 126   1+2+6 = 9
12.411 / 2 = 6.2055   6+2+0+5+5 = 18   1+8 = 9
9523431 / 2 = 4761715.5
            4+7+6+1+7+1+5+5= 36   3+6 =9

DR9 divided by five = DR9

18 / 5 = 3.6 =9
27 / 5 = 5.4 = 9
63 / 5 = 12.6 = 9
171 / 5 = 34.2 = 9
432 / 5 = 86.4 = 18 = 9
621 / 5 = 124.2 = 9

9 or any DR9 divided by any number equals 9 or DR9*

9 / 2 = 4.5   4+5 = 9
9 / 5 = 1.8   1+8 = 9
9 / 8 = 1.125   1+1+2+5 = 9
45 / 4 = 11.25   1+1+2+5 = 9
135 / 5 = 27   1+3+5=9 2+7 = 9   (135=9)
135 / 8 = 16.875   1+6+8+7+5 = 27 2+7=9
567 / 7 = 81   8 + 1 = 9   (567=18=9)
657 / 5 = 131.4   6+5+7 =18   1+8 =9   1+3+1+4=9
567 / 4 = 141.75   1+4+1+7+5 = 18   1+8=9
45 / 2 = 22.5   2+2+5 = 9   27 = 9
714285 / 5 = 142857   7+1+4+2+8+5=27 = 9
1+4+2+8+5+7 = 27 = 9
*- Except 3 and 6: 9/3=3   9/6=1.5

Number Divided by 9 (N/9)

Any digit divided by 9 yields a repeating sequence of only one number that is equal to 'n' or its digital sum.

1/9 = 0.11111111111...
2/9 = 0.22222222222...

3/9 = 0.33333333333...
4/9 = 0.44444444444...
5/9 = 0.55555555555...
6/9 = 0.66666666666...
7/9 = 0.77777777777...
8/9 = 0.88888888888...
9/9 = 1
10/9 = 1.1111111111...
11/9 = 1.2222222222...
12/9 = 1.3333333333...
13/9 = 1.4444444444...
14/9 = 1.5555555555...
15/9 = 1.6666666666...
16/9 = 1.7777777777...
17/9 = 1.8888888888...
18/9 = 2
19/9 = 2.1111111111...
20/9 = 2.2222222222...
21/9 = 2.3333333333...
22/9 = 2.4444444444...
23/9 = 2.5555555555...
24/9 = 2.6666666666...
25/9 = 2.7777777777...
26/9 = 2.8888888788...
27/9 = 3
28/9 = 3.1111111111...
29/9 = 3.2222222222...
30/9 = 3.3333333333...
36/9 = 4
40/9 = 4.4444444444...
41/9 = 4.5555555555...
42/9 = 4.6666666666...
45/9 = 5
50/9 = 5.5555555555...
53/9 = 5.8888888888...
54/9 = 6

The Simple Complexity of Number Nine

56/9 = 6.2222222222...
60/9 = 6.6666666666...
66/9 = 7.3333333333...
70/9 = 7.7777777777...
72/9 = 8
73/9 = 8.1111111111...
74/9 = 8.2222222222...
80/9 = 8.8888888888...
81/9 = 9
88/9 = 9.7777777777...
90/9 = 10
91/9 = 10.1111111111...
92/9 = 10.2222222222...
93/9 = 10.3333333333...
94/9 = 10.4444444444...
95/9 = 10.5555555555...
96/9 = 10.6666666666...
97/9 = 10.7777777777...
98/9 = 10.8888888888...
99/9 = 11
100/9 = 11.1111111111...
101/9 = 11.222222222...
102/9 = 11.333333333...
103/9 = 11.444444444...
104/9 = 11.555555555...
105/9 = 11.666666666...
106/9 = 11.777777777...
107/9 = 11.888888888...
108/9 = 12
109/9 = 12.111111111...
110/9 = 12.222222222...
111/9 = 12.333333333...

Triples of same number divided by 9
111 / 9 = 12.333...
222 / 9 = 24.333... (This equals twice the previous number).

333 / 9 = 37
444 / 9 = 49.333...
555 / 9 = 61.666...
666 / 9 = 74 (equals twice 333/9)
777 / 9 = 86.333...
888 / 9 = 98.666...(equals twice 444/9)
999 / 9 = 111 (equals three times 333/9)

1/1089 (1+0+8+9 = 18 = 9)

1/1089 = 0.000918273645546372 81... (The decimal expansion is a sequence of numbers of the **9** times table: 9, 18 (9x2), 27 (9x3), 36 (9x4), 45 (9x5), 54 (9x6), 63 (9x7), 72 (9x8), 81 (9x9)...)

7 or DR7 divided by 9
7 / 9 = 0.7777777...
16 / 9 = 1.7777777...
25 / 9 = 2.7777777...
34 / 9 = 3.7777777...
43 / 9 = 4.7777777...
52 / 9 = 5.7777777...
61 / 9 = 6.7777777...
70 / 9 = 7.7777777...

The division of any number by the amount of 9s corresponding to its number of digits, the number is turned into a repeating decimal.

Examples:

37 / 99 = 0.373737373737...
58 / 99 = 0.585858585858...
274 / 999 = 0.274274274274...
414 / 999 = 0.414414414414414414...

Any number divided by 9 that has a remainder that equals the sum of that number.

Example: 485 divided by 9 = 53.8, so 8 is the remainder and the sum of 4+8+5=17 and 7+1=8.

Repeating Nines, DS9 or DR9

Any number divided by 11, except repetition in a two digit number[*], the decimal numbers will have a digital root of nine.

7 / 11 = 0.63636363...   3+6 = 9
25 / 11 = 2.27272727...   2+7 = 9
634 / 11 = 57.636363...   6+3 = 9
5641 / 11 = 512.818181...   8+1 = 9
54123 / 11 = 4920.272727...   2+7 = 9
259814 / 11 = 23619.454545... 4+5 = 9
(*)- 99/11=9   55/11=5

1/11 = 0.0909090... (90 repeats)
2/11 = 0.18181818... (18 repeats. 1+8=9)
3/11 = 0.27272727... (27 repeats. 2+7=9)
4/11 = 0.36363636... (36 repeats. 3+6=9)
5/11 = 0.45454545... (45 repeats. 4+5=9)
6/11 = 0.54545454... (54 repeats. 5+4=9)
7/11 = 0.63636363... (63 repeats. 6+3=9)
8/11 = 0.72727272... (72 repeats. 7+2=9)
9/11 = 0.81818181... (81 repeats. 8+1=9)
10/11= 0.9090909... (90 repeats. 9+0=9)
12/11 = 1.09090909... (09 repeats. 0+9=9)
13/11 = 1.18181818... (18 repeats. 1+8=9)
21/11 = 1.90909090... (90 repeats. 9+0=9)
30/11 = 2.72727272... (72 repeats. 7+2=9)
52/11 = 4.72727272... (72 repeats. 7+2=9)
263/11 = 23.90909090... (90 repeats. 9+0=9)
4213/11 = 383 (!)...
4231/11 = 384.636363...(63 repeats. 6+3=9)
1234/11 = 112.181818... (18 repeats. 1+8=9)
12345/11 = 1122.272727...(27 repeats. 2+7=9)
123456/11 = 11223.2727... (27 repeats. 2+7=9)

1234567/11= 112233.3636... (36 repeats. 3+6=9)
12345678/11=1122334.3636...(36 repeats. 3+6=9)
123456789/11=11223344.4545.(45 repeats. 4+5=9)

Note the mirror image of the decimal run in the first 10 numbers:

09 18 27 36 45 * 54 63 72 81 90!!!

1/11 = 0.09090
12/11 = 1.090909
123/11 = 11.181818
1234/11 = 112.181818
12345/11 = 1122.272727
123456/11 = 11223.272727
1234567/11 = 112233.363636
12345678/11 = 1122334.363636
123456789/11 = 11223344.454545

DR9 / DR9 = Repeating decimals
13320 / 594 = 22.4242424242...
23310 / 495 = 47.0909090909...
59400 / 13320 = 4.4594594594 (4+5+9=18=9)
6174 / 369 = 20.6341463414 (6+3+4+1+4=18=9)
7641 / 369 = 20.7073170731 (7+7+3+1=18=9)
7614 / 369 = 20.6341463414 (6+3+4+1+4=18=9)
7461 / 369 = 20.2195121951 (2+1+9+5+1=18=9)

9 or DR9 / 5 = 9 or DR9
9/5 = 1.8   1+8 = 9
18/5 = 3.6   3+6 = 9
27/5 = 5.4   5+4 = 9
135/5 = 27   2+7 = 9
432/5 = 86.4 8+6+4 = 18 = 9
621/5 = 124.2   1+2+4+2 = 9
4212/5 = 842.4   8+4+2+4 = 18 = 9
23634/5 = 4726.8   4+7+2+6+8 = 27 = 9

Division of runs of 1 to 9
(Each fraction has all the digits from 0 to 9)

97524/10836 = 9
95823/10647 = 9
95742/10638 = 9
75249/08361 = 9
58239/06471 = 9
57429/06381 = 9!!!

Division of runs of 1 to 8
18273645 / 9 = 2030405.0
45362718 / 9 = 5040302.0
(2030405 & 5040302)
54637281 / 9 = 6070809.0
81726354 / 9 = 9080706.0
(6070809 & 9080706)

Number with 33 in
If a number has '33' in it, then:

1. It is divisible by 9
2. The digital sum of the number will also be divisible by 9.
   36: = 4x3x3. Thirty-six is therefore divisible by 9 giving 4. 3+6 = 9
   279: = 31x3x3. 2+7+9 = 18 = 9.

Quick division by 99

Since 99 is a 2 digit number, then omit 2 digits from the number to be divided (the dividend), and then add 1 to it.

This rule is only applicable if:

1. The dividend is evenly divisible by the number consisting of only 9 (Divisor).
2. The number of digits of the divisor (9's) is half or more than the number of digits of the dividend i.e. dividing a 4-digits number by

99 or 999, a 6-digits number by 999 or 9999 or 99999, an 8-digits number by four 9's or more ... and so on.

6237 / 99:
Remove 2 digits from the dividend 6237 to become 62; then add 1 to become 63.
63 is the answer! 6237/9 = 63
1188 / 99:
Remove 2 digits. 1188 becomes 11.
Add 1, 11 become 12.
Therefore, 1188 / 99 = 12

Quick division by 999
Follow the same rule used in the previous entry. But as 999 is a three-digit number then three numbers will have to be omitted from the dividend.

22977 / 999:
Remove 3 digits from the dividend. 22977 become 22; add 1 to become 23. Then 23 is the answer of the division (22977/999).
61938 / 999:
Remove 3 digits and 61938 become 61.
Add 1, and 61 become 62.

Therefore, 61938 / 999 = 62!

The same rule applies to other numbers. Division by four nines will require the omission of four digits, division by 5 nines will require the omission of five nines, etc.

Quick division by 99999
7846421535 / 99999:
7846421535 become 78464 after the removal of five digits.
Then add one to get the answer 78465!

Number Trees of 9

9+9... = 18

$$9$$
$$9 = 18$$
$$99 = 108$$
$$999 = 1008$$
$$9999 = 10008$$
$$99999 = 100008$$
$$999999 = 1000008$$
$$9999999 = 10000008$$
$$99999999 = 100000008$$
$$999999999 = 1000000008$$
$$9999999999 = 10000000008$$

9x9... = 81

$$9$$
$$9 = 81$$
$$99 = 891$$
$$999 = 8991$$
$$9999 = 89991$$
$$99999 = 899991$$
$$999999 = 8999991$$
$$9999999 = 89999991$$
$$99999999 = 899999991$$
$$999999999 = 8999999991$$

9-9... = 0

$$9$$
$$9 = \mathbf{0}$$
$$99 = 9\mathbf{0}$$
$$999 = 99\mathbf{0}$$
$$9999 = 999\mathbf{0}$$
$$99999 = 9999\mathbf{0}$$
$$999999 = 99999\mathbf{0}$$
$$9999999 = 999999\mathbf{0}$$
$$99999999 = 9999999\mathbf{0}$$
$$999999999 = 99999999\mathbf{0}$$

9/9... = 1

$$9$$
$$9 = \mathbf{1}$$
$$99 = \mathbf{11}$$
$$999 = \mathbf{111}$$
$$9999 = \mathbf{1111}$$
$$99999 = \mathbf{11111}$$
$$999999 = \mathbf{111111}$$
$$9999999 = \mathbf{1111111}$$
$$99999999 = \mathbf{11111111}$$
$$999999999 = \mathbf{111111111}$$

1 x 9 + N...

$$0 \times 9 + 1 = 1$$
$$1 \times 9 + 2 = 11$$
$$12 \times 9 + 3 = 111$$
$$123 \times 9 + 4 = 1111$$
$$1234 \times 9 + 5 = 11111$$
$$12345 \times 9 + 6 = 111111$$
$$123456 \times 9 + 7 = 1111111$$
$$1234567 \times 9 + 8 = 11111111$$
$$12345678 \times 9 + 9 = 111111111$$
$$123456789 \times 9 + 10 = 1111111111$$

Notice the number of 1s in the sum is equal to the number added in the formula, i.e. in the second line the number added is 3, and the number of 1s in the result is 3, and so on.

Special Numbers (I)

1. 124578
2. 142857
3. 875421
4. 12345678
5. 27
6. 45
7. 126
8. 153
9. 234
10. 315
11. 360
12. 369
13. 441
14. 567
15. 621
16. 666
17. 711
18. 854
19. 1089
20. 1239
21. 5040
22. 6174
23. 0123456789
24. Perfect numbers
25. The nine tables
26. Nine times a sequence minus the sequence

(1) - 124578…124578…124578…
(1 to 10 excluding 3, 6 & 9)

This is a bewildering number with remarkable flexibility, which I stumbled upon along my journey in the mazes of numbers...

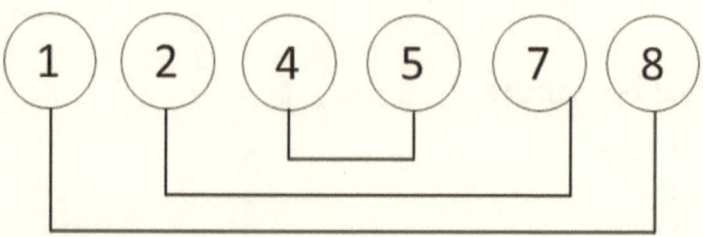

The connected numbers are equal to nine.

Multiplication

x 2 = 249156.   2+4+9+1+5+6 = 27 = 9
x 3 = 373734.   3+7+3+7+3+4 = 27 = 9
x 4 = 498312.   4+9+8+3+1+2 = 27 = 9
x 5 = 622890.   6+2+2+8+9+0 = 27 = 9
x 6 = 747468.   7+4+7+4+6+8 = 36 = 9
x 7 = 872046.   8+7+2+0+4+6 = 27 = 9
x 8 = 996624.   9+9+6+6+2+4 = 36 = 9
x 9 =1121202.   1+1+2+1+2+0+2   = 9

Reverse Multiplication

124578 x 875421 = 109058197338 = 54 = 9
714285 x 582417 = 4160117265845 = 45 = 9

124578 x N.decimal = 999,999
714285 x 1.4 = 999,999
428571 x 2.33 = 999,999
285714 x 3.5 = 999,999
857142 x 1.166 = 999,999
571428 x 1.75 = 999,999

Division

124578/2 = 62289   = 27 = 9
      /3 = 41526   = 18 = 9

/4 = 13842   = 18 = 9
/5 = 24915.6   = 27 = 9
/7 = 17796.8…
/8 = 15572.25 = 27 = 9
/9 = 13842   = 18 = 9

123456789 / 124578 = 990.999927756…

The first 9 numbers:
    9+9+0+9+9+9+9+2+7 =36 = 9
And, the first two numbers: 9+9 = 18 = 9
The first three numbers: 9+9+9 = 27 = 9
The first four numbers: 9+9+9+9 = 36 = 9
The first five numbers: 9+9+9+9+9 = 45 = 9
The first six numbers: 9+9+9+9+9+9 = 54 = 9

124578 / 999,999
1245 / 999999 = 0.124578124578…

Division of 9 or DR9 by 7
Note the recurrence of number 124578 in various sequences, such as 285714, 571428, 857142, 142857, 428571, 714285…

9 / 7 = 1.2857142857142857142…
18 / 7 = 2.5714285714285714285…
27 / 7 = 3.857142857142857142…
36 / 7 = 5.142857142857142857…
45 / 7 = 6.428571428571428571…
54 / 7 = 7.714285714285714285…
63 / 7 = 9
72 / 7 = 10.285714285714285714…
81 / 7 = 11.571428571428571428…
90 / 7 = 12.857142857142857142…
108 / 7 = 15.428571428571428571…
702 / 7 = 100.285714285714285714…
900 / 7 = 128.571428571428571428…

Exceptions: 504/7 = 72

Addition

1) - 124578 + Next number sequence
124578+245781 = 370359 = 27 = 9
124578+457812 = 582390 = 27 = 9
124578+578124 = 702702 = 18 = 9
124578+781245 = 905823 = 27 = 9
124578+812457 = 937035 = 27 = 9
124578 +875421 = 999999 = 54 = 9

2) - 124578 + run of sequence
124578
245781
457812
578124
781245
<u>812457</u>
2999997   = 6 nines (five nines + 27) = 54 = 9
2999997 / 5 = 599999.4 = 6 nines = 54 = 9
2999997 / 7 = 428571 (!) = 27 = 9
2999997 / 9 = 333333 = 18 = 9

3) - Reverse Addition
124578+875421 = 999999 = 54 = 9

4) - 124578 and its variant, e.g. 142857, always yield the same numbers but in different arrangement; except for numbers starting with 8:

142857 + 285714 = 428571 = 27 = 9
142857 + 428571 = 571428 = 27 = 9
142857 + 571425 = 714285 = 27 = 9
142857 + 714285 = 857142 = 27 = 9
142857 + 857142 = 999999 = 54 = 9
142857 + 812745 = 955602 = 27 = 9

The Simple Complexity of Number Nine

5) - Subtraction
875421 - 124578 = 750843 = 27 = 9

6) - 285714
x 2 = 571428   = 27 = 9
x 3 = 857142   = 27 = 9
x 4 = **1**142856 (1+6= missing 7) = 27 = 9
x 5 = **1**428570 (1+0=1)   = 27 = 9
x 6 = **1**714284 (1+4= missing 5) = 27 = 9
x 7 = 1810998 = 36 = 9
x 8 = **2**285712 (2+2= missing 4) = 27 = 9
x 9 = **2**571426 (6+2= missing 8) = 27 = 9

7) - Any number divided by 7 will have 124578 in the decimal position, in any sequence, in the result

(Except 49, 56, 77, 98 and 91)
1)-11/7 = 1.571428571
2)-12/7 = 1.714285714
3)-19/7 = 2.714285714
4)-23/7 = 3.285714286
5)-39/7 = 5.571428571
6)-45/7 = 6.714285714
7)-51/7 = 7.285714285
8)-55/7 = 7.857142857
9)-67/7 = 9.571428571
10)-78/7 = 11.14285714
11)-89/7 = 12.71428571
12)-90/7 = 12.85714286
13)-92/7 = 13.14285714
14)-99/7 = 14.14285714
15)-152/7 = 21.7142857142
16)-652/7 = 93.1428571428
17)-850/7 = 121.428571428
18)-925/7 = 132.142857142
19)-1255/7 = 179.285714285

8) - Division of all numbers
123456789 / 124578 = 990.999927756...
9+9+0+9+9+9+9+2+7+7+5+6 = 117 = 9

9) - 124578 x 124578
= 15519678084
= 54 = 9

10) - 124578 x 875421
= 1090588197338
= 72 = 9

11) - 142857 x 5
= 714285!

12) - 428571 / 3
= 142857!

13) - 142857 Addition

```
  142857   142857   142857   142857   142857
+ 285714  +428571  +571428  +714285  +857142
  428571   571428   714285   857142   999999!
```

(2) – 142857... 142857...142857...

This number, which is a variant of the previous number, **124578**, is sometimes referred to as the 'phoenix number' because its first six multiples are anagrams of it...

142857 x 1 = **1**42857
142857 x 2 = **2**85714
142857 x 3 = **4**28571
142857 x 4 = **5**71428
142857 x 5 = **7**14285
142857 x 6 = **8**57142

- The descending numbers of the first column of the resultants is **124578...**

- The ascending numbers of the fourth column of the resultant is **124578...**

Multiplication of variations of runs of 142857 by 7 equals DR9:

142857 x 7 = 999999 (six 9s = 54 = 9)
425571 x 7 = 2999997 (2+7=9 six 9s...)
285714 x 7 = 1999998 (1+8=9 six 9s...)
857142 x 7 = 5999994 (5+4=9 six 9s...)
571428 x 7 = 3999996 (3+6=9 six 9s...)
714285 x 7 = 4999995 (4+5=9 six 9s...)

Addition of variations of runs of 142857 equals DR9:

142857 + 285714 = 428571 = 27 = 9
142857 + 428571 = 571428
142857 + 571428 = 714285
285714 + 428571 = 714285!
285714 + 142857 = 428571
285714 + 714285 = 999999
428571 + 571428 = 999999!

9 or DR9 divided by 7 will always result with the number in any order, except 63:

9 / 7 = 1.2857142...
15 / 7 = 2.1428571...
27 / 7 = 3.8571428...
36 / 7 = 5.1428571...
45 / 7 = 6.4285714...
54 / 7 = 7.7142857...
63 / 7 = 9
72 / 7 = 10.285714...
81 / 7 = 11.571428...
90 / 7 = 12.857142...
99 / 7 = 14.142857...
108 / 7 = 15.4285714...

207 / 7 = 29.5714228…
306 / 7 = 43.7142857…
702 / 7 = 100.285714285…
801 / 7 = 114.428571428…
900 / 7 = 128.571428571…

Five and its increments of 5, results in a variety of 714285:

5 / 7 = 0.7142857…
10 / 7 = 1.4285714…
15 / 7 = 2.1428571…
20 / 7 = 2.8571428…
25 / 7 = 3.5714285…
30 / 7 = 4.2857142…
35 / 7 = 5.0
40 / 7 = 5.7142857…
45 / 7 = 6.4285714…
50 / 7 = 7.1428571…
95 / 7 = 13.571428…

360 divided by 7 and sequences of 7
360 / 7 = 51.42857 142857 142857 142857 1428571…
360 / 14 = 25.7 142857 142857 142857 142857…
360 / 21 = 17.142857 142857 142857 142857…
360 / 28 = 12.857142857 142857 142857 142857…
360 / 35 = 10.2857 142857 142857 142857 142857…
360 / 42 = 8.57 142857 142857 142857 142857…
360 / 56 = 6.42857 142857 142857 142857 142857…
360 / 63 = 5.7 142857 142857 142857 142857…

142857: 1+4+2+8+5+7 = 27 = 9!
360 / 49 = 7.34693877551020408163265306122 44897959183673469…
49 is the only number in this sequence that does not follow the others in yielding 142857. It is to be noted that 49 is 7x7…
360 / 77 = 4.67532 467532 467532 467532 4675324… This sequence includes all the numbers between 2 and 7 – 234567!

The Simple Complexity of Number Nine

2+3+4+5+6+7 = 27 = 9!
2x3x4x5x6x7 = 5040 = 9!

(3) – 875421... 875421...875421...

This number is the reverse of the earlier number 124578, i.e. in a descending order.

		8	7	5	4	2	1	
A	x2	1	7	5	0	8	4	2 =27=9
B	x3	2	6	2	6	2	6	3 =45=9
C	x4	3	5	0	1	6	8	4 =27=9
D	x5	4	3	7	7	1	0	5 =27=9
E	x6	5	2	5	2	5	2	6 =27=9
F	x7	6	1	2	7	9	4	7 =36=9
G	x8	7	0	0	3	3	6	8 =27=9
H	x9	7	8	7	8	7	8	9 =54=9
		35	32	28	34	41	38	44 =252=9
		8	5	10	7	5	11	8 =54=9
		8	5	1	7	5	2	8 =36=9

Note:

- The shaded rows:

2626263, 5252526 and 7878789

```
 2626263         2626263 = 45 = 9
+5252526         5252526 = 27 = 9
 7878789!!!     +7878789 = 54 = 9
                15757578 = 45 = 9
```

- Addition of rows:

A + B = D	B + C = F
A + C = E	B + D = G
A + D = F	B + E = H
A + E = G	…………..
A + F = H	C + D = H

(4) – 12345678...12345678...12345678...

18273645 / 9 = 2030405
54637281 / 9 = 6070809

(18273645 + 54637281 = 72910926 = 36 = 9)
81726354 / 9 = 9080706

1 2 3 4 5 6 7 8

1234 = 10 = 1
<u>5678</u> = 26 = <u>8</u>
6912 = 18 = 9
1234 = 10 = 1
<u>8765</u> = 26 = <u>8</u>
9999   36 = 9

(5) – 27…27…27…
(2+7=9)
27 x 2 = 54 (= 9)
27 x 3 = 72 (= 9)
27 x 4 = 108 (= 9)
27 x 5 = 135 (= 9)
27 x 6 = 162 (= 9)
27 x 7 = 189 (= 18 = 9)
27 x 8 = 216 (= 9)
27 x 9 = 243 (= 9)

27 x 3, x 3, x3…
27 x 3 = 81 (= 9)
81 x 3 = 243 (= 9)
243 x 3 = 729 (=18 =9)
729 x 3 = 2,187 (=18= 9)
2187 x 3 = 6,561 (=18=9)
6561 x 3 = 19,683 (=27=9)
19683 x 3 = 59,049 (=27=9)

59049 x 3 = 177,147 (=27=9)
177147 x 3 = 531,441 (=18=9)
531441 x 3 = 1,594,323 (=27=9)
1594323 x 3 = 4,782,969 (=45=9)
4782969 x 3 = 14,348,907 (=36=9)
14348907 x 3 = 43,046,721 (=27=9)
43046721 x 3 = 129,140,163 (=27=9)
129140163 x 3 = 387,420,489 (=45=9)
387420489 x 3 = 1,161,261,467 (=36=9)

The same applies to any other number, whose digital root is equal to 9, e.g.:
36 x 3 = 108 (=9)
108 x 3 = 324 (=9)
324 x 3 = 972 (=18=9)
45 x 3 = 135 (=9)
135 x 3 = 405 (=9)
405 x 3 = 1215 (=9)
81 x 3 = 243 (=9)
243 x 3 = 729 (=18=9)

27 is equal to the sum of the digits of its cube:

27^3 = 19683
1+9+6+8+3 = 27!

(6) – 45…45…45…
45 is divisible only by 1, 3, 5 and 9.
4+5 = 9   1+3+5+9 = 18 = 9

(7) – 126…126…126…
(162…261…216…612…621)
126 x 126 = 15876 = 27 = 9
126 x 621 = 78246 = 27 = 9
126 x 216 = 27216 = 18 = 9
621 / 126 = 4.9<u>285714</u> <u>285714</u>… (285714= 27 = 9)
126 + 621 = 495 =   18 = 9

621 + 216 = 837 = 18 = 9
621 + 126 = 747 = 18 = 9
621 - 126 = 495 = 18 = 9
621 - 216 = 405   = 9
621 - 162 = 459 = 18 = 9
126 x 9 =   1134   = 9
126 x 99 =   12474 = 18 = 9
126 x 999 = 125874 = 27 = 9
216 x 999 = 215784 = 27 = 9

(8) – 153…153…153…
- It is the smallest number that can be expressed as the sum of the cubes of all its digits: $153 = 1^3+5^3+3^3$. This is called an Armstrong number. Another example of such a number: $371 = 3^3+7^3+1^3$.
- The sum of the digits is a perfect square: 1+5+3=9!
- 153 + 351 = 504 = 9
- 504 x 504 (504^2) = 254016   2+5+4+0+1+6 = 18 = 9
- 504^2 = 288 x 882   2+8+8 = 18 = 9
- 153 is the 17th triangular number. Its reverse 351 is also a triangular. Hence, 153 is a reversible triangular number!
  The first few triangle numbers: 1, 3, 6, 10, 15, 21, 28, 36, 45, 55, 66, 78, 91, 105, 120, 136, **153**, 171, 190, 210, 231, 253, 276, 300, 325, **351**, 378, 406, 435, 465, 496, 528, 561, 595, 630, 666, 703, 741, 780, 820, 861, 903, 946, 990, 1035, 1081, 1128, 1176, 1225, 1275, 1326, 1378, 1431, 1485, 1540, 1596, 1653, 1711, 1770, 1830, 1891, 1953, 2016, 2080…
- It is also the 9th hexagonal number.

1, 6, 15, 28, 45, 66, 91, 120, **153**, 190, 231, 276, 325, 378, 435, 496, 561, 630, 703, 780, 861, 946, 1035, 1128, 1225, 1326, 1431, 1540, 1653, 1770, 1891, 2016, 2145, 2278, 2415, 2556, 2701, 2850, 3003, 3160, 3321, 3486, 3655, 3828, 4005, 4186, 4371, 4560…

- Number 153 is divisible by the sum of its own digits (DS) and so it is a Harshad number:
  153/1+5+3 =17.
- 153 is the product of two numbers which formed its own: 153 = 3 x 51
- Since 153 = 100 + 28 + 25 it is considered to represent harmonisation of contrasts: 100 represents a square (10th square number), 28 represents a triangle (7th triangular number) and 25 represents a circle. 25 is also the tum of the 3rd and 4th square numbers = 9+16=25.
- 153153 is the smallest odd abundant number ending in 3.
- 153 is equal to the sum of factorials of number from 1 to 5 – 153 = 1! + 2! + 3! + 4! + 5!
- The Bible tells of Jesus and the Apostles going fishing and catching exactly **153** fish.
- The sum of aliquot divisors of 153 is also a perfect square: 1 + 3 + 9 + 17 + 51 = 81 = 9
- 153 x n = DR9
  153 x 5 = 765 = 18 = 9
  153 x 12 = 1836 = 18 = 9
  153 x 648 = 99144 = 27 = 9
  153 x 935 = 143055 = 18 = 9
- 153 – 9 or DS9 or DR9 = DR9
  153 – 9 = 144 = 9
  153 – 36 = 117 = 9
  153 – 72 = 63 = 9
  153 – 144 = 9
  153 – 342 = -189 = 18 = 9
  153 – 900 = -747 = 18 = 9
  153 – 378 = -225 = 9
  153 – 531 = -378 = 18 = 9
  153 – 351 = -198 = 18 = 9
- Addition of highest and lowest number from these 135 number will result in 666
  135 + 531 = 666 = 18 = 9
  153 + 513 = 666

315 + 351 = <u>666</u>
1998 = 27 = 9
- 153 x (12345678 + 87654321) + 153 = 15,299,999,847 + 153 = 153,000,000,000
- The square root of 153, 12.369, is the number of full moons in one year.
- 65359477124183 x 153 = 9999999999999999!
- 513 = $5^3 + 1^3 + 3^3$ = 125 + 1 + 27 = 153!

(9) – 234…234…234…
234 x 234 =   54756 = 27 = 9
234 x 432 = 101088 = 18 = 9
324 x 342 = 110808 = 18 = 9
234 + 432 =   666 = 18 = 9
234 + 234 =   468 = 18 = 9
234 + 324 =   558 = 18 = 9
234 + 342 =   576 = 18 = 9
243 + 342 =   567 = 18 = 9
234 + 243 =   477 = 18 = 9
423 + 324 =   747 = 18 = 9
432 + 432 = 864 = 18 = 9
432 / 234 = 1.<u>846153</u> <u>846153</u>… (846153=27=9)
432 / 243 = 1.7777777777777…

(10) – 315…315…315…
315 x 2 = 630 = 9
315 x 51 = 16065 = 18 = 9
315 x 99 = 3185 = 18 = 9
315 + 513 = 828 = 18 = 9
513 – 315 = 198 = 18 = 9

(11) – 360…360…360…
360 x 2 = 720 = 9
360 x 22 = 7920 = 18 = 9
360 x 3 = 1080 = 9
360 x 33 = 11880 =18 =9
360 x 4 = 1440 = 9

360 x 44 = 15840 = 18 = 9
360 x 5 = 1800 = 9
360 x 55 = 19800 = 18 = 9
360 x 6 = 2160 = 9
360 x 666 = 239760 = 27 = 9
360 x 7 = 2520 = 9
360 x 77 = 27720 = 18 = 9
360 x 777 = 279720 = 27 = 9
360 x 8 = 2880 = 18 = 9
360 x 88 = 31680 = 18 = 9
360 x 9 = 3240 = 9
360 x 9999 = 3599640 = 36 = 9

360 / 2 = 180 = 9
360 / 4 = 90 = 9
360 / 5 = 72 = 9
360 / 7 = 51.42857142857.... 142857 = 27 = 9
360 / 14 = 25.7142857142857... 142857 = 27 = 9
360 / 16 = 22.5 = 9
360 / 21 = 17.142857142857... 142857 = 27 = 9
360 / 25 = 14.4 = 9
360 / 32 = 11.25 = 9
360 / 35 = 10.285714 285714 2857...
142857 = 27 = 9

Note that the sequence 142857 does not contain the numbers 3, 6 or 9…

Divide 360 by the number of days in the year:

360 / 364.5 = 0.9876543209876543209…!

(12) – 369…369…369…
$\sqrt{9} = 3$
$\sqrt{36} = 6$
$\sqrt{81} = 9$
369 / 3 = 123

(13) – 441...441...441...
441 x 144 = 63504 = 18 = 9
414 x 144 = 59616 = 18 = 9
441 + 144 = 585 = 18 = 9
441 + 414 = 855 = 18 = 9
414 + 414 = 828 = 18 = 9
441 – 144 = 297 = 18 = 9
414 – 144 = 270 = 9
441 – 414 = 27 = 9

(14) – 567...567...567...
567^2 = 321,489

The numbers used in this equation are all the numbers from 1 to 9, and the digital sum of 321489 is 27 which has a digital root of 9!

(15) – 621...621...621...
621 x 621 = 385641 = 27 = 9
621 x 126 = 78246 = 27 = 9
621 – 126 = 495 = 18 = 9
621 – 216 = 405 = 9
621 + 621 = 1242 = 9
621 + 126 = 747 = 18 = 9

(16) – 666...666...666...
666 x 1 = 666 = 18 = 9
666 x 2 = 1332   = 9
666 x 3 = 1998 = 27 = 9
666 x 4 = 2664* = 18 = 9
666 x 5 = 3330   = 9
666 x 6 = 3996 = 27 = 9
666 x 7 = 4662* = 18 = 9
666 x 8 = 5328 = 18 = 9
666 x 9 = 5994 = 27 = 9
(*) – 2664 and 4662 are a mirror image!

666 x 22 = 14652 = 18 = 9
666 x 33 = 21978 = 27 = 9
666 x 44 = 29304 = 18 = 9
666 x 55 = 36630 = 18 = 9
666 x 66 = 43956* = 27 = 9
666 x 77 = 51282 = 18 = 9
666 x 88 = 58608 = 27 = 9
666 x 99 = 65934* = 27 = 9
(*) - 43956 and 65934 are a mirror image!

666 x 222 = 147852* = 3 nines = 27 = 9
666 x 333 = 221778 = 3 nines = 27 = 9
666 x 444 = 295704 = 3 nines = 27 = 9
666 x 555 = 369630 = 3 nines = 27 = 9
666 x 666 = 443556 = 3 nines = 27 = 9
666 x 777 = 517482* = 3 nines = 27 = 9
666 x 888 = 591408 = 3 nines = 27 = 9
666 x 999 = 665334 = 3 nines = 27 = 9
DS in all equals 27...
(*) - 147852, the 'Mysterious Number'...

666 x 9 = 5994 = 3 nines = 27
666 x 18 = 11988 = 3 nines = 27
666 x 27 = 17982 = 3 nines = 27
666 x 36 = 23976 = 3 nines = 27
666 x 45 = 29970 = 3 nines = 27
666 x 54 = 35964* = 3 nines = 27
666 x 63 = 41958 = 3 nines = 27
666 x 72 = 47952 = 3 nines = 27
666 x 81 = 53946* = 3 nines = 27
(*) - Same numbers but in a different order...

 (17) – 711...711...711...
711 + 117 = 828 = 18 = 9
711 + 171 = 882 = 18 = 9
711 + 711 = 1422 = 9

117 + 117 = 234 = 9
171 + 171 = 342 = 9
711 x 711 = 505521 = 18 = 9
711 − 117 = 594 = 18 = 9

(18) − 854…854…854…
854^2 = 729,316
Again, the numbers used in this equation are all the numbers from 1 to 9.

(19) − 1089…1089…1089…
1000 / 1089 = 0.91827364554
The decimal expansion is a sequence of the nine times table: 9, 18, 27, 36, 45, and 54.

(20) − 1239…1239…1239…
123456789. 123+234+345+456+567+678+789 = 3192. (Ascending sequence **1239**)
1239 x 9 = 11151 = 9
1293 x 9 = 11637 = 18 = 9
1329 x 9 = 11961 =18 = 9
1392 x 9 = 12528 = 18 = 9
1923 x 9 = 17307 = 18 = 9
<u>1932</u> x 9 = 17388 = 27 = 9
9108 = 18 = 9
2193 x 9 = 19737 = 27 = 9
2139 x 9 = 19251 = 18 = 9
2319 x 9 = 20871 = 18 = 9
2391 x 9 = 21519 = 18 = 9
2913 x 9 = 26217 = 18 = 9
<u>2931</u> x 9 = 26379 = 27 = 9
14886 = 27 = 9

3129 x 9 = 28161 = 18 = 9
3192 x 9 = 28728 = 27 = 9
3219 x 9 = 28971 = 27 = 9
3291 x 9 = 29619 = 27 = 9

3912 x 9 = 35208 = 18 = 9
3921 x 9 = 35289 = 27 = 9
20664 = 18 = 9

9123 x 9 = 82107 = 18 = 9
9132 x 9 = 82188 = 27 = 9
9213 x 9 = 82917 = 27 = 9
9231 x 9 = 83079 = 27 = 9
9312 x 9 = 83808 = 27 = 9
9321 x 9 = 83889 = 36 = 9
55332 = 18 = 9

(21) – 5040…5040…5040…
1x2x3x4x5x6x7 = 5040 = 9
7x8x9x10 = 5040 = 9

(22) – 6174…6174…6174…
This number is known as 'Kaprekar's constant' after the Indian mathematician D. R. Kaprekar.

Following the steps below (Kaprekar's routine), the result will always yield 6174:

1. Take any four-digit number, using at least two different digits. Leading zeros are allowed.
2. Arrange the digits in descending and then in ascending order to get two four-digit numbers, adding leading zeros if needed.
3. Subtract the smaller number from the bigger number.
4. Repeat step 2.
5. After a few repetitions of the steps, the result will be 6174!
   (6174: 6+1+7+4 = 18 = 9)

Example: 6421

6421 – 1246 = 5175
7551 – 1557 = 5355
5553 – 3555 = 1998

9981 − 1899 = 8082
8820 − 0288 = 8532
8532 − 2358 = 6174
7164 − 1467 = 6174!

Note that this follows the same formula mentioned earlier under 'Subtraction', regardless of the arrangements of the numbers: **DR9 − DR9 = DR9**

Examples:

7416 − 6741 = 675 = 18 = 9
6174 − 4761 = 1413 = 9
4617 − 1764 = 2853 = 18 = 9
8352 − 5823 = 2529 = 18 = 9
6390 − 639 = 5751 = 18 = 9
6174 − 774 = 5400 = 9
8352 − 234 = 8118 = 18 = 9

(23) − 0123456789 variations times five
0246913578 x 5 = 1234567890!
1975308642 x 5 = 9876543210!

Perfect Numbers

The perfect numbers are extremely rare.

There is only one between 1 and 10, namely, 6; one between 10 and 100, namely, 28; one between 100 and 1,000, namely, 496; and one between 1,000 and 10,000, namely, 8,128.

```
   6
  28
 946
8128
9108  →  9+1+0+8 = 18  →  1+8 = 9
```

The nine tables

0 1 2 3 4 5 6 7 8 9
9 8 7 6 5 4 3 2 1 0

This is a mnemonic trick to remember the nine tables.
9x9=81 9x8=72 9x7=63 9x6=54 9x5=45
9x4=36 9x3=27 9x2=18

9x9 = 81   18 = 2x9
9x8 = 72   27 = 3x9
9x7 = 63   36 = 4x9
9x6 = 54   45 = 5x9
9x5 = 45   54 = 6x9
9x4 = 36   63 = 7x9
9x3 = 27   72 = 8x9
9x2 = 18   81 = 9x9

Nine times digital sequence minus digital sequence equals reversed digital sequence
(9 x 123,456,789) - 123,456,789 = 987,654,321

Central Nine!

987654321 − 123456789 =
8641**9**7532
(8+6+4+1+9+7+5+3+2= 45 = 9)

Special Numbers (II)

1- Fibonacci numbers
2- Lucas numbers
3- Golden ratio
4- Prime numbers
5- Plastic constant
6- Cubic numbers

(1) - Fibonacci Numbers (Series)

Fibonacci, the greatest European mathematician of the middle ages, was born in Pisa, Italy; hence his full name was Leonardo of Pisa, or Leonardo Pisano in Italian. Leonardo's father, Guglielmo Bonacci, was a kind of customs officer in the present-day Algerian town of Béjaïa, formerly known as Bugia or Bougie, where wax candles were exported to France. Candles are still called "bougies" in French.

Leonardo grew up with a North African education under the Moors and later travelled extensively around the Mediterranean coast. He would have met with many merchants and learned of their systems of doing arithmetic. He soon realised the many advantages of the "Hindu-Arabic" system over all the others.

He was one of the first people to introduce the Hindu-Arabic number system into Europe - the positional system we use today - based on ten digits with its decimal point and a symbol for zero:

1 2 3 4 5 6 7 8 9 0

His book on how to do arithmetic in the decimal system, called 'Liber Abaci' (meaning *Book of the Abacus* or *Book of Calculating*) completed in 1202 persuaded many European mathematicians of his day to use this "new" system.

The book describes (in Latin) the rules we all now learn at elementary school for adding numbers, subtracting, multiplying and dividing, together with many problems to illustrate the methods.

Fibonacci says his book *Liber Abaci* (the first edition was dated 1202) that he had studied the "nine Indian figures" and their arithmetic as used in various countries around the Mediterranean and wrote about them to make their use more commonly understood in his native Italy. So he probably merely included the "rabbit problem" from one of his contacts and did not invent either the problem or the series of numbers which now bear his name.

It was the French mathematician Edouard Lucas (1842-1891) who gave the name Fibonacci numbers to this series and found many other important applications as well as having the series of numbers that are closely related to the Fibonacci numbers - the Lucas Numbers: 2, 1, 3, 4, 7, 11, 18, 29, 47, ... named after him.

The 'Fibonacci Sequence' is a series of numbers:

0, 1, 1, 2, 3, 5, 8, 13, 21, 34... The next number is found by adding up the two numbers before it.

The series are both additive and multiplicative: Each number is the sum of the two previous ones, and each number approximates the previous number multiplied by the golden section.

<u>0, 1, 1, 2, 3, 5, 8, 13, 21</u>, 34...The first 9 numbers have a DS of 54 and a DR of 9: 5+4=9

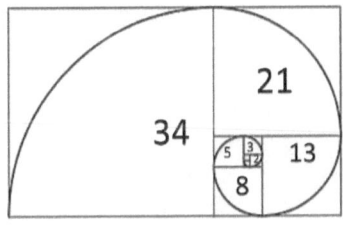
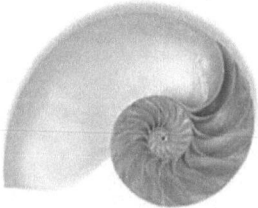

The first 300 Fibonacci numbers

n : Fn (Fibonacci number)
0 : 0
1 : 1
2 : 1
3 : 2
4 : 3
5 : 5
6 : 8 = 2^3
7 : 13
8 : 21 = 3 x 7
9 : 34 = 2 x 17
10 : 55 = 5 x 11
11 : 89
12: 144 = 24 x 32 (1+4+4=9).
13 : 233
14 : 377 = 13 x 29
15 : 610 = 2 x 5 x 61
16 : 987 = 3 x 7 x 47
17 : 1597
18 : 2584 = 2^3 x 17 x 19
19 : 4181 = 37 x 113
20 : 6765 = 3 x 5 x 11 x 41
21 : 10946 = 2 x 13 x 421
22 : 17711 = 89 x 199
23 : 28657
24: 46368 = 2^5 x 3^2 x 7 x 23 (4+6+3+6+8=27=9).
25 : 75025 = 5^2 x 3001
26 : 121393 = 233 x 521
27 : 196418 = 2 x 17 x 53 x 109
28 : 317811 = 3 x 13 x 29 x 281
29 : 514229
30 : 832040 = 2^3 x 5 x 11 x 31 x 61
31 : 1346269 = 557 x 2417
32 : 2178309 = 3 x 7 x 47 x 2207
33 : 3524578 = 2 x 89 x 19801

The Simple Complexity of Number Nine

34 : 5702887 = 1597 x 3571
35 : 9227465 = 5 x 13 x 141961
36: 14930352 = 2^4 x 3^3 x 17 x 19 x 107 (1+4+9+3+3+5+2=27=9).
37 : 24157817 = 73 x 149 x 2221
38 : 39088169 = 37 x 113 x 9349
39 : 63245986 = 2 x 233 x 135721
40 : 102334155 = 3 x 5 x 7 x 11 x 41 x 2161
41 : 165580141 = 2789 x 59369
42 : 267914296 = 2^3 x 13 x 29 x 211 x 421
43 : 433494437
44 : 701408733 = 3 x 43 x 89 x 199 x 307
45 : 1134903170 = 2 x 5 x 17 x 61 x 109441
46 : 1836311903 = 139 x 461 x 28657
47 : 2971215073
48: 4807526976 = 2^6 x 3^2 x 7 x 23 x 47 x 1103 (4+8+7+5+2+6+9+7+6=54=9).
49 : 7778742049 = 13 x 97 x 6168709
50 : 12586269025 = 5^2 x 11 x 101 x 151 x 3001
51 : 20365011074 = 2 x 1597 x 6376021
52 : 32951280099 = 3 x 233 x 521 x 90481
53 : 53316291173 = 953 x 55945741
54 : 86267571272 = 2^3 x 17 x 19 x 53 x 109 x 5779
55 : 139583862445 = 5 x 89 x 661 x 474541
56 : 225851433717 = 3 x 7^2 x 13 x 29 x 281 x 14503
57 : 365435296162 = 2 x 37 x 113 x 797 x 54833
58 : 591286729879 = 59 x 19489 x 514229
59 : 956722026041 = 353 x 2710260697
60: 1548008755920 = 2^4 x 3^2 x 5 x 11 x 31 x 41 x 61 x 2521 (1+5+4+8+8+7+5+5+9+2=54=9).
61 : 2504730781961 = 4513 x 555003497
62 : 4052739537881 = 557 x 2417 x 3010349
63 : 6557470319842 = 2 x 13 x 17 x 421 x 35239681
64 : 10610209857723 = 3 x 7 x 47 x 1087 x 2207 x 4481
65 : 17167680177565 = 5 x 233 x 14736206161
66 : 27777890035288 = 2^3 x 89 x 199 x 9901 x 19801
67 : 44945570212853 = 269 x 116849 x 1429913
68 : 72723460248141 = 3 x 67 x 1597 x 3571 x 63443

69 : 117669030460994 = 2 x 137 x 829 x 18077 x 28657

70 : 190392490709135 = 5 x 11 x 13 x 29 x 71 x 911
    x 141961

71 : 308061521170129 = 6673 x 46165371073

72: 498454011879264 = 2^5 x 3^3 x 7 x 17 x 19 x 23 x 107 x 103681 (4+9 +8+4+5+4+1+1+8+7+9+2+6+4=72=9).

73 : 806515533049393 = 9375829 x 86020717

74 : 1304969544928657 = 73 x 149 x 2221 x 54018521

75 : 2111485077978050 = 2 x 5^2 x 61 x 3001 x 230686501

76 : 3416454622906707 = 3 x 37 x 113 x 9349 x 29134601

77 : 5527939700884757 = 13 x 89 x 988681 x 4832521

78 : 8944394323791464 = 2^3 x 79 x 233 x 521 x 859
    x 135721

79 : 14472334024676221 = 157 x 92180471494753

80 : 23416728348467685 = 3 x 5 x 7 x 11 x 41 x 47 x 1601
    x 2161 x 3041

81 : 37889062373143906 = 2 x 17 x 53 x 109 x 2269
    x 4373 x 19441

82 : 61305790721611591 = 2789 x 59369 x 370248451

83 : 99194853094755497

84: 160500643816367088 = 2^4 x 3^2 x 13 x 29 x 83 x 211 x 281 x 421 x 1427 (1+6+5+6+4+3+8+1+6+3+6+7+8+8=72=9)

85 : 259695496911122585 = 5 x 1597 x 9521 x 3415914041

86 : 420196140727489673 = 6709 x 144481 x 433494437

87 : 679891637638612258 = 2 x 173 x 514229
    x 3821263937

88 : 1100087778366101931 = 3 x 7 x 43 x 89 x 199 x 263 x 307 x 881 x 967

89 : 1779979416004714189 = 1069 x 1665088321800481

90 : 2880067194370816120 = 2^3 x 5 x 11 x 17 x 19 x 31 x 61
    x 181 x 541 x 109441

91 : 4660046610375530309 = 13^2 x 233 x 741469
    x 159607993

92 : 7540113804746346429 = 3 x 139 x 461 x 4969 x 28657
    x 275449

93 : 12200160415121876738 = 2 x 557 x 2417
    x 4531100550901

94 : 19740274219868223167 = 2971215073 x 6643838879

95 : 31940434634990099905 = 5 x 37 x 113 x 761 x 29641
   x 67735001

96: 51680708854858323072 = 2^7 x 3^2 x 7 x 23 x 47 x 769 x 1103 x 2207 x 3167 (5+1+6+8+7+8+8+5+4+8+5+8+3+2+3+7+2=90=9).

97 : 83621143489848422977 = 193 x 389 x 3084989
   x 361040209

98 : 135301852344706746049 = 13 x 29 x 97 x 6168709
   x 599786069

99 : 218922995834555169026 = 2 x 17 x 89 x 197 x 19801
   x 18546805133

100 : 354224848179261915075 = 3 x 5^2 x 11 x 41 x 101
   x 151 x 401 x 3001 x 570601

101 : 573147844013817084101 = 743519377
   x 770857978613

102 : 927372692193078999176 = 2^3 x 919 x 1597
   x 3469 x3571 x 6376021

103 : 1500520536206896083277 = 519121 x 5644193
   x 512119709

104 : 2427893228399975082453 = 3 x 7 x 103 x 233 x 521
   x 90481 x 102193207

105 : 3928413764606871165730 = 2 x 5 x 13 x 61 x 421
   x 141961 x 8288823481

106 : 6356306993006846248183 = 953 x 55945741
   x 119218851371

107 : 10284720757613717413913
   = 1247833 x 8242065050061761

108: 16641027750620563662096 = 2^4 x 3^4 x 17 x 19 x 53 x 107 x 109 x 5779 x 11128427 (1+6+6+4+1+2+7+7+5+6+2+5+6+3+6+6+2+9+6=90=9)

109 : 26925748508234281076009 = 827728777
   x 32529675488417

110 : 43566776258854844738105 = 5 x 11^2 x 89 x 199
   x 331 x 661 x 39161 x 474541

111 : 70492524767089125814114 = 2 x 73 x 149 x 2221
   x 1459000305513721

112 : 114059301025943970552219 = 3 x 7^2 x 13 x 29 x 47
   x 281 x 14503 x 10745088481

113 : 18455182579303309636633
 = 677 x 27260240146681402729
114 : 29861112681897706691855² = 2³ x 37 x 113 x 229 x 797 x 9349 x 54833 x 95419
115 : 483162952612010163284885 = 5 x 1381 x 28657 x 2441738887963981
116 : 78177407943098723020343⁷ = 3 x 59 x 347 x 19489 x 514229 x 1270083883
117 : 126493703204299739348832² = 2 x 17 x 233 x 29717 x 135721 x 39589685693
118 : 20467111114739846236917⁵⁹ = 353 x 709 x 8969 x 336419 x 2710260697
119 : 3311648143516982017180081 = 13 x 1597 x 159512939815855788121
120: 535835925499096664087184⁰ = 2⁵ x 3² x 5 x 7 x 11 x 23 x 31 x 41 x 61 x 241 x 2161 x 2521 x 20641 (5+3+5+8+3+5+5+9+2+5+4 +9+9+9+6+6+6+4+8+7+1+8+4=126=9)
121: 86700073985079486580519²¹ = 89 x 974158134663814455960⁸⁹
122: 140283666534989152989237⁶¹ = 4513 x 555003497 x 5600748293801
123: 226983740520068639569756⁸² = 2 x 2789 x 59369 x 68541957733949701
124: 367267407055057792558994⁴³ = 3 x 557 x 2417 x 3010349 x 3020733700601
125: 594251147575126432128751²⁵ = 5³ x 3001 x 1584141679640457000⁰¹
126: 9615185546301842246877456⁸ = 2³ x 13 x 17 x 19 x 29 x 211 x 421 x 1009 x 31249 x 35239681
127: 15557697022053106568164969³ = 27941 x 556805304822773221007³
128: 25172882568354948815042426¹ = 3 x 7 x 47 x 127 x 1087 x 2207 x 4481 x 186812208641
129: 40730579590408055383207395⁴ = 2 x257 x 5417 x 8513 x 39639893 x 433494537
130: 65903462158763004198249821⁵ = 5 x 11 x 131 x 233 x 521 x 2081 x 24571 x 14736206161
131: 106634041749171059581457216⁹

**132: 17253750390793406377970703840384 = 2^4 x 3^2 x 43 x 89 x 199 x 307
x 9901 x 19801 x 261399601 (1+7+2+5+3+7+5+3+9+7+9+3+4+6+3
+7+7+9+7+7+3+8+4=126=9)**

133: 27917154565710512336116425553 = 13 x 37 x 113
x 3457 x 42293 x 351301301942501

134: 4517090495650391871408712937 = 269 x 4021
x 116849 x 1429913 x 24994118449

135: 73088059522214431050203555490 = 2 x 5 x 17 x 53
x 61 x 109 x 109441 x 1114769954367361

136: 11825896447871834976429068427 = 3 x 7 x 67
x 1597 x 3571 x 63443 x 23230657239121

137: 191347024000932780814494423917

138: 309605988479651130578784922344 = 2^3 x 137 x 139
x 461 x 691 x 829 x 18077 x 28657 x 1485571

139: 500953012480583911393279166261
= 277 x 2114537501 x 855267229376896093

140: 810559000960235041972064086055 = 3 x 5 x 11 x 13
x 29 x 41 x 71 x 281 x 911 x 141961 x 12317523121

141: 1311512013440818953365343244866 = 2
x 108289 x 1435097 x 142017737 x 2971215073

142: 2122071014401053995337407333471 = 6673 x
46165371073 x 688846502588399

143: 34335830278418729487027505058337 = 89 x 233
x 8581 x 19295841537568504966621

**144: 555565404224292694404015791808 = 2^6 x 3^3 x 7 x 17 x 19 x 23
x 47 x 107 x 1103 x 103681 x 10749957121 (5+5+5+5+6+5+4+4+
2+2+4+2+9+2+6+9+4+4+4+1+5+7+9+1+8+8=126=9)**

145: 898923707008479989274290850145 = 5 x 514229
x 349619996930737079890201

146: 1454489111232772683678306641953 = 151549
x 9375829 x 86020717 x 11899937029

147: 2353412818241252672952597492098 = 2 x 13 x 97
x 293 x 421 x 3529 x 6168709 x 347502052673

148: 3807901929474025366309041334051 = 3 x 73 x 149
x 2221 x 11987 x 54018521 x 81143477963

149: 6161314747715278029583501626149 = 110557
x 162709 x 4000949 x 85607646594577

150: 996921667718930338621440576020 = 2^3 x 5^2 x 11 x 31 x 61 x 101 x 151 x 3001 x 12301 x 18451 x 230686501

151: 1613053142490458141579790738634 9 = 5737 x 2811666624 525811646469915877

152: 260997481020938848020123131465 49 = 3 x 7 x 37 x 113 x 9349 x 29134601 x 1091346396980401

153: 4223027952699846621781022053289 8 = 2 x 17^2 x 1597 x 6376021 x 7175323114950564593

154: 683300276290923510198225336794 47 = 13 x 29 x 89 x 199 x 229769 x 988681 x 4832521 x 9321929

155: 110560307156090817237632754212 345 = 5 x 557 x 2417 x 21701 x 12370533881 x 61182778621

156: 17889033478518316825745528789179 2 = 2^4 x 3^2 x 79 x 233 x 521 x 859 x 90481 x 135721 x 12280217041 (1+7+8+8+9+3+3+4+7+8+5+1+8+3+1+6+8+2+5+7+4+5+2+8+7+8+9+1+7+9+2=171=9)

157: 289450641941273985495088042104 137 = 313 x 11617 x 7636481 x 10424204306491346737

158: 468340976726457153752543329995 929 = 157 x 92180471494753 x 32361122672259149

159: 757791618667731139247631372100 066 = 2 x 317 x 953 x 55945741 x 97639037 x 229602768949

160 : 122613259539418829300017470209 5995 = 3 x 5 x 7 x 11 x 41 x 47 x 1601 x 2161 x 2207 x 3041 x 23725145626561

161: 198392421406191943224780607419 6061 = 13 x 8693 x 28657 x 612606107755058997065597

162: 321005680945610772524798077629 2056 = 2^3 x 17 x 19 x 53 x 109 x 2269 x 3079 x 4373 x 5779 x 19441 x 62650261

163: 519398102351802715749578685048 8117 = 977 x 4892609 x 33365519393 x 32566223208133

164: 840403783297413488274376762678 0173 = 3 x 163 x 2789 x 59369 x 800483 x 350207569 x 370248451

165: 135980188564921620402395544772 68290 = 2 x 5 x 61 x 89 x 661 x 19801 x 86461 x 474541 x 518101 x 900241

166: 220020566894662969229833221040 48463 = 35761381 x 6202401259 x 99194853094755497

167: 356000755459584589632228765813 16753 = 18104700793 x 1966344318693345608565721

The Simple Complexity of Number Nine

168: 57602132235424755886206198685365216 = 2^5 x 3^2 x 7^2 x 13 x 23 x 29 x 83 x 167 x 211 x 281 x 421 x 1427 x 14503 x 65740583 (5+7+6+2+1+3+2+2+3+5+4+2+4+7+5+5+8+8+6+2+6+1+9+8+6+8+5+3+6+5+2+1+6=153=9)

169: 9320220778138321484942907526668196 9 = 233 x 337 x 89909 x 104600155609 x 126213229732669

170: 15080434001680797073563527395204718 5 = 5 x 11 x 1597 x 3571 x 9521 x 1158551 x 12760031 x 3415914041

171: 24400654779819118558506434921872915 4 = 2 x 17 x 37 x 113 x 797 x 6841 x 54833 x 5741461760879844361

172: 39481088781499915632069962317077633 9 = 3 x 6709 x 144481 x 433494437 x 313195711516578281

173: 63881743561319034190576397238950549 3 = 1639343785721 x 38967874900762927 1532733

174: 10336283234281894982264635955602818 32 = 2^3 x 59 x 173 x 349 x 19489 x 514229 x 947104099 x 3821263937

175: 16724457590413798401322275679497873 25 = 5^2 x 13 x 701 x 3001 x 141961 x 17231203730201189308301

176: 27060740824695693383586911635100691 57 = 3 x 7 x 43 x 47 x 89 x 199 x 263 x 307 x 881 x 967 x 93058241 x 562418561

177: 43785198415109491784909187314598564 82 = 2 x 353 x 2191261 x 805134061 x 1297027681 x 2710260697

178: 70845939239805185168496098949699256 39 = 179 x 1069 x 1665088321800481 x 22235502640988369

179: 11463113765491467695340528626429782 121 = 21481 x 156089 x 3418816640903898929534613769

180: 185477076894719862121901385213997077 60 = 2^4 x 3^3 x 5 x 11 x 17 x 19 x 31 x 41 x 61 x 107 x 181 x 541 x 2521 x 109441 x 10783342081 (1+8+5+4+7+7+7+6+8+9+4+7+1+9+8+6+2+1+2+1+9+1+3+8+5+2+1+3+9+9+7+7+7+6=180=9)

181: 300108214549634539075306671478294898 81 = 8689 x 422453 x 8175789237238547574551461093

182: 485585291444354401197208056692291976 41 = 13^2 x 29 x 233 x 521 x 741469 x 159607993 x 689667151970161

183: 785693505993988940272514728170586875 22 = 2 x 1097 x 4513 x 555003497 x 14297347971975757800833

184: 127127879743834334146972278486287885163 = 3 x 7 x 139 x 461 x 4969 x 28657 x 253367 x 275449 x 9506372193863

185: 2056972303432332281742237513033465727685 = 5 x 73 x 149 x 2221 x 17029455131913055569070971618161

186: 3328251100870675623211960297896344578848 = 2^3 x 557 x 2417 x 63799 x 3010349 x 35510749 x 4531100550901

187: 5385223404303007904954197810929810305335 = 89 x 373 x 1597 x 10 15780730596343409910503491703

188: 87134745051736835281661581088261548838 = 3 x 563 x 5641 x 2971215073 x 6643838879 x 4632894751907

189: 1409869790947669143312035591975596518914 = 2 x 13 x 17 x 53 x 109 x 421 x 38933 x 35239681 x 9559219503167350377

190: 22812172414650374961286514028582120072955 = 5 x 11 x 37 x 113 x 191 x 761 x 9349 x 29641 x 41611 x 67735001 x 87382901

191: 3691087032412706639440686994833808526209 = 4870723671313 x 757810806256989128439975793

192: 5972304273877744135569338397692020533504 = 2^8 x 3^2 x 7 x 23 x 47 x 769 x 1087 x 1103 x 2207 x 3167 x 4481 x 11862575248703 (5+9+7+2+3+4+2+7+3+8+7+7+7+4+4+1+3+5+5+ 6+9+3+3+8+3+9+7+6+9+2+2+5+3+3+5+4=180=9)

193: 9663391306290450775010025392525829059713 = 9465278929 x 1020930432032326933976826008497

194: 15635695580168194910579363790217849593217 = 193 x 389 x 3299 x 3084989 x 361040209 x 56678557502141579

195: 25299086886458645685589389182743678652930 = 2 x 5 x 61 x 233 x 135721 x 14736206161 x 88999250837499877681

196: 40934782466626840596168752972961528246147 = 3 x 13 x 29 x 97 x 281 x 5881 x 6168709 x 599786069 x 61025309469041

197: 66233869353085486281758142155705206899077 = 15761 x 25795969 x 227150265697 x 717185107125886549

198: 10716865181971232687792689512866673545224 = 2^3 x 17 x 19 x 89 x 197 x 199 x 991 x 2179 x 9901 x 19801 x 1513909 x 18546805133

199: 17340252117297813159685037284371942044301 = 397 x 43678216 9201002048261171378550055269633

200: 28057117299251014003761193241303867718925 = 3 x 5^2 x 7 x 11 x 41 x 101 x 151 x 401 x 2161 x 3001 x 570601 x 9125201 x 5738108801

201: 45397369416530795319729696969741061923382 = 2 x 269 x 505026070439624716931599021 x 1429913 x 116849

The Simple Complexity of Number Nine

202: 7345448671578180932349089021104492964233511 = 809 x 7879 x 743519377 x 770857978613 x 201062946718741

203: 11885185613231260464322058718078599156571771 = 13 x 1217 x 514229 x 56470541 x 258698270065673399465953

204: **192306342848094413966711477391830921208052811 = 2^4 x 3^2 x 67 x 409 x 919 x 1597 x 3469 x 3571 x 63443 x 6376021 x 66265118449**

(1+9+2+3+6+3+4+2+8+4+8+9+4+4+1+3+9+6+6+7+1+1+4 +7+7+3+9+1+8+3+9+2+1+2+8+5+2+8=180)

205: 3111581989804070186099320645726169127737705 = 5 x 821 x 2789 x 59369 x 125598581 x 36448117857891321536401

206: 50346454182850143257664354196444783398182331 = 619 x 1031 x 519121 x 5644193 x 512119709 x 5257480026438961

207: 8146227408089084511865756065370647467555938 = 2 x 17 x 137 x 829 x 18077 x 28657 x 40723531557736276012221964811

208: 13180872826374098837632191485015125807374171 = 3 x 7 x 47 x 103 x 233 x 521 x 3329 x 90481 x 102193207 x 106513889 x 325759201

209: 21327100234463183349497947550385773274930109 = 37 x 89 x 113 x 57314120955051297736679165379998262001

210: 34507973060837282187130139035400899082304280 = 2^3 x 5 x 11 x 13 x 29 x 31 x 61 x 71 x 211 x 421 x 911 x 21211 x 141961 x 767131 x 8288823481

211: 55835073295300465536628086585786672357234389 = 22504837 x 38490197 x 800972881 x 80475423858449593021

212: 90343046356137747723758225621187571439538669 = 3 x 953 x 1483 x 2969 x 55945741 x 119218851371 x 1076012367720403

213: 146178119651438213260386312206974243796773058 = 2 x 1277 x 6673 x 46165371073 x 185790722054921374395775013

214: 236521166007575960984144537828161815236311727 = 1247833 x 47927441 x 479836483312919 x 8242065050061761

215: 382699285659014174244530850035136059033084785 = 5 x 433494437 x 2607553541 x 67712817361580804952011621

216: **619220451666590135228675387863297874269396512 = 25 x 34 x 7 x 17 x 19 x 23 x 53 x 107 x 109 x 5779 x 6263 x 103681 x 11128427 x 177962167367**

(6+1+9+2+2+0+4+5+1+6+6+6+5+9+0+1+3+5+2+2+8+6+7+5+3+8+7+8+ 6+3+2+9+7+8+7+4+2+6+9+3+9+6+5+1+2=216 =9)

217: 1001919737325604309473206237898433933302481297 = 13 x 433 x 557 x 2417 x 44269 x 217221773 x 2191174861 x 6274653314021

218: 1621140188992194444701881625761731807571877809 = 128621 x 788071 x 827728777 x 593985111211 x 32529675488417

219: 2623059926317798754175087863660165740874359106 = 2 x 123953 x 4139537 x 9375829 x 86020717 x 3169251245945843761

220: 4244200115309993198876969489421897548446236915 = 3 x 5 x 11^2 x 41 x 43 x 89 x 199 x 307 x 331 x 661 x 39161 x 474541 x59996854928656801

221: 6867260041627791953052057353082063289320596021 = 233 x 1597 x 203572412497 x 90657498718024645326392940793

222: 11111460156937785151929026842503960837766832936 = 2^3 x 73 x 149 x 2221 x 4441 x 146521 x 1121101 x 54018521 x 1459000305513721

223: 17978720198565577104981084195586024127087428957 = 4013 x 108377 x 251534189 x 164344610046410138896156070813

224: 29090180355503362256910111038089984964854261893 = 3 x 7^2 x 13 x 29 x 47 x 223 x 281 x 449 x 2207 x 14503 x 10745088481 x1154149773784223

225: 47068900554068939361891195233676009091941690850 = 2 x 5^2 x 17 x 61 x 3001 x 109441 x 230686501 x 11981661982050957053616001

226: 76159080909572301618801306271765994056795952743 = 677 x 272602401466814027129 x 412670427844921037470771

227: 123227981463641240980692501505442003148737643593 = 23609 x 5219534137983025159078847113619467285727377

228: 1993870623732135425994938077772079972055335 96336 = 24 x 32 x 37 x 113 x 227 x 229 x 797 x 9349 x 26449 x 54833 x 95419 x 29134601 x 212067587
(1+9+9+3+8+7+6+2+3+7+3+2+1+3+5+4+2+5+9+9+4+9+3+8+7+7 +7+7+2+7+9+9+7+2+5+5+3+3+5+9+6+3+3+6=234=9)

229 : 322615043836854783580186309282650000354271239929 = 457 x 2749 x 40487201 x 132605449901 x 47831560297620361798553

230: 522002106210068326179680117059857997559804836265 = 5 x 11 x 139 x 461 x 1151 x 1381 x 5981 x 28657 x 324301 x 686551 x 2441738887963981

231: 844617150046923109759866426342507997914076076194 = 2 x 13 x 89 x 421 x 19801 x 988681 x 4832521 x 9164259601748159235188401

THE SIMPLE COMPLEXITY OF NUMBER NINE

232: 1366619256256991435939546543402365995473880912459 = 3 x 7 x 59 x 347 x 19489 x 299281 x 514229 x 1270083883 x834428410879506721

233: 2211236406303914545699412969744873993387956988653 = 139801 x 25047390419633 x 631484089583693149557829547141

234: 357785566256090598163895951314723998861837901112 = 2^3 x 17 x 19 x 79 x 233 x 521 x 859 x 29717 x 135721 x 39589685693 x1052645985555841

235: 5789092068864820527338372482892113982249794889765 = 5 x 2971215073 x 389678426275593986752662955603693114561

236: 9366947731425726508977331996039353971111632790877 = 3 x 353 x 709 x 8969 x 336419 x 15247723 x 2710260697 x100049587197598387

237: 15156039800290547036315704478931467953361427680642 = 2 x 157 x 1668481 x 40762577 x 92180471494753 x 7698999052751136773

238: 24522987531716273545293036474970821924473060471519 = 13 x 29 x 239 x 1597 x 3571 x 10711 x 27932732439809 x 159512939815855788121

239: 39679027332006820581608740953902289877834488152161 = 10037 x 62141 x 2228536579597318057 x 28546908862296149233369

240: 642020148637230941269017774288731118023075486 23680 = 26 x 32 x 5 x 7 x 11 x 23 x 31 x 41 x 47 x 61 x 241 x 1103 x 1601 x 2161 x 2521 x 3041 x 20641 x 23735900452321
6+4+2+0+2+0+1+4+8+6+3+7+2+3+0+9+4+1+2+6+9+0+1+7+7+7+4+2+8+8+7+3+1+1+1+8+0+2+3+0+7+5+4+8+6+2+3+6+8+0=19 8=18=9

241: 103881042195729914708510518382775401680142036775841 = 11042621 x 7005329677 x 1342874889289644763267952824739273

242: 168083057059453008835412295811648513482449585399521 = 89 x 199 x 97415813466381445596089 x 97420733208491869044199

243: 271964099255182923543922814194423915162591622175362 = 2 x 17 x 53 x 109 x 2269 x 4373 x 19441 x 448607550257 x16000411124306403070561

244: 44004715631463592379335110006072428645041207574883 = 3 x 4513 x 19763 x 21291929 x 555003497 x 5600748293801 x24848660119363

245: 712011255569818855923257924200496343807632829750245 = 5 x 13 x 97 x 141961 x 6168709 x128955073914024460192651484843195641

246: 1152058411884454788302593034206568772452674037325128 = 2^3 × 2789 × 59369 × 4767481 × 370248451 × 7188487771 × 68541957733949701

247: 1864069667454273644225850958407065116260306867075373 = 37 × 113 × 233 × 409100738617 × 4677306043367904676926312147 328153

248: 3016128079338728432528443992613633888712980904400501 = 3 × 7 × 557 × 743 × 2417 × 467729 × 3010349 × 3020733700601 × 33758740830460183

249: 4880197746793002076754294951020699004973287771475874 = 2 × 1033043205255409 × 99194853094755497 × 23812215284009787769

250: 7896325826131730509282738943634332893686268675876375 = 5^3 × 11 × 101 × 151 × 251 × 3001 × 112128001 × 28143378001 × 158414167964045700001

251: 12776523572924732586037033894655031898659556447352249 = 582416774750273 × 21937080329465122026187124199656961913

252: 20672849399056463095319772838289364792345825 123 228624 = 2^4 × 3^3 × 13 × 17 × 19 × 29 × 83 × 107 × 211 × 281 × 421 × 1009 × 1427 × 31249 × 1461601 × 35239681 × 764940961

253: 33449372971981195681356806732944396691005381570580873 = 89 × 28657 × 4322114369 × 2201228236641589 × 1378497303338047612061

254: 54122222371037658776676579571233761483351206693809497 = 509 × 5081 × 27941 × 487681 × 13822681 × 19954241 × 5568053048227732210073

255: 87571595343018854458033386304178158174356588264390370 = 2 × 5 × 61 × 1597 × 9521 × 6376021 × 3415914041 × 20778644396941 × 20862774425341

256: 141693817714056513234709965875411919657707794958199867 = 3 × 7 × 47 × 127 × 1087 × 2207 × 4481 × 119809 × 186812208641 × 4698167634523379875583

257: 229265413057075367692743352179590077832064383222590237 = 5653 × 32971978671645905645521 × 12300267 21719313471360714649

258: 370959230771131880927453318055001997489772178180790104 = 2^3 × 257 × 5417 × 6709 × 8513 × 144481 × 308311 × 39639893 × 433494437 × 761882591401

259: 60022464382820724862019667023459207532183656140338 0341 =
13 x 73 x 149 x 1553 x 2221 x 404656773793 x3041266742295771
985148799223649

260 : 9711838745993391295476499882895940728116087395841 70445 =
3 x 5 x 11 x 41 x 131 x 233 x 521 x 2081 x 3121 x 24571 x 90481 x
14736206161 x 42426476041450801

261: 15714085184275463781678466585241861481334453009875 50786 =
2 x 17 x 173 x 2089 x 20357 x 36017 x 40193 x 322073 x 514229 x
3821263937 x 6857029027549

262: 254259239302688550771549664681378022094505404057 1721
231 = 1049 x 414988698461 x 5477332620091 x 10663404174
91710595814572169

263: 41140009114544318858833433053379663690784993415592 72017 =
4733 x 93629 x92836229646390194235291216984425664630 89
390281

**264: 6656593304481317393598839952151746590023 5533 82130
993248 = 2^5 x 3^2 x 7 x 23 x 43 x 89 x 199 x 263 x 307 x 881 x
967 x 5281 x 9901 x 19801 x 66529 x 152204449 x 261399601**

265: 10770594215935749279482183257489712959102052723690 265265
= 5 x 953 x 15901 x 55945741 x 2741218753681 x926918599
457468125920827581

266: 17427187520417066673081023209641459549125606105821 2585
13 = 13 x 29 x 37 x 113 x 3457 x 9349 x 42293 x 10694421739
x2152958650459 x 351301301942501

267: 281977817363528159525632064671311725082276588295 1152
3778 = 2 x 1069 x 122887425153289 x 1665088321800481
x6445587734970304 2877309

268: 45624969256769882625644229676772632057353264935332 7822
91 = 3 x 269 x 4021 x 6163 x 116849 x 1429913 x 24994118449
x201912469249 x 2705622682163

269: 73822750993122698578207436143903804565580923764844
306069 = 5381 x 2517975182669813 x 32170944747810641
x169360439829648789853

270: 11944772024989258120385166582067643662293418870017 7088360
= 2^3 x 5 x 11 x 17 x 19 x 31 x 53 x 61 x 109 x 181 x 271 x 541 x 811 x
5779 x 42391 x 109441 x 119611 x 1114769954367361

271: 19327047124301527978205910196458024118851511246502139 4429 = 449187076348273 x430267212525867121951740619093594938 058573

272: 312718191492907860985910767785256677811449301165198 4827 89 = 3 x 7 x 47 x 67 x 1597 x 3571 x 63443 x 23230657239121 x562627837283291940137654881

273: 505988662735923140767969869749836918999964413630219877218 = 2 x 13 x 13 x 233 x 421 x 135721 x 640457 x 741469 x 159607993 x1483547330343905886515273

274: 81870685422883100175388063753509359681141371479541 83 60007 = 541721291 x 78982487870939058281 x 191347 024000932780 81449423917

275: 132469551696475414252185050728493051581137812842563823 72 25 = 5^2 x 89 x 661 x 3001 x 474541 x 7239101 x 15806979101 x5527278404454199535821801

276: 21434023711935851442757311448200241126227918 4 3221056597232 = 2^4 x 3^2 x 137 x 139 x 461 x 691 x 829 x 4969 x 16561 x 18077 x 28657 x 162563 x 275449 x 1485571 x 1043766587

277: 3468097888158339286797581652104954628434169971646694834457 = 505471005740691524853293621 x6861121308187330908986 328104917

278: 56115002593519244310733127969249787410569618148677514316 89 = 277 x 30859 x 253279129 x 2114537501 x 14331800109223159 x 85526722937689093

279: 9079598147510263717870894449029933369491131786514446266146 = 2 x 17 x 557 x 2417 x 11717 x 4531100550901 x 594960058508093x 6279830532252706321

280: 146910984068621881489442072459549121105480936013821976978 35 = 3 x 5 x 7^2 x 11 x 13 x 29 x 41 x 71 x 281 x 911 x 2161 x 14503 x 141961 x 12317523121 x 118021448662479038881

281: 237706965543724518668151016949848454800392253878966439639 81 = 174221 x 119468273 x114205973520041784262049438829 32 15303693455057

282: 384617949612346400157593089409397575905873189892788416 61816 = 2^3 x 108289 x 1435097 x 79099591 x 142017737 x 2971215073 x 6643838879 x 139509555271

The Simple Complexity of Number Nine

283: 62232491515607091882574410635924603070626544377175485625797 = 10753 x 825229 x 15791401 x444111888848805843163235784298630863264881

284: 100694286476841731898333719576864360661213863366454327287613 = 3 x 283 x 569 x 6673 x 2820403 x 9799987 x 35537616083 x 46165371073 x 688846502588399

285: 162926777992448823780908130212788963731840407743629812913410 = 2 x 5 x 37 x 61 x 113 x 761 x 797 x 29641 x 54833 x 67735001 x956734616715046328502480330601

286: 263621064469290555679241849789653324393054271110084140201023 = 89 x 199 x 233 x 521 x 8581 x 1957099 x 2120119 x1784714380021 x 1929584153756850496621

287: 426547842461739379460149980002442288124894678853713953114433 = 13 x 2789 x 59369 x 198160071001853267796700692507490184570501064382201

288: 69016890693102993513939182979209561251794894 9963798093315456 = 2^7 x 3^3 x 7 x 17 x 19 x 23 x 47 x 107 x 769 x 1103 x 2207 x 3167 x 103681 x 10749957121 x 115561578124838522881

289: 1116716749392769314599541809794537900642843628817512046429889 = 577 x 1597 x 1733 x 98837 x 101232653 x 106205194357 x658078658277725444483848541

290: 1806885656323799249738933639586633513160792578781310139745345 = 5 x 11 x 59 x 19489 x 514229 x 120196353941 x 1322154751061x 349619996930737079890201

291: 2923602405716568564338475449381171413803636207598822186175234 = 2 x 193 x 389 x 3084989 x 361040209 x76674415738994499773 x 227993117754975870677

292: 4730488062040367814077409088967804926964428786380132325920579 = 3 x 29201 x 151549 x 9375829 x 86020717 x 11899937029 x37125857850184727260788881

293: 7654090467756936378415884538348976340768064993978954512095813 = 64390759997 x11886939163497285252295209896447615 5238134997314729

294: 12384578529797304192493293627316781267732493780359086838016392 = 2^3 x 13 x 29 x 97 x 211 x 293 x 421 x 3529 x 65269 x 620929 x 6168709 x 8844991 x 599786069 x 347502052673

295: 20038668997554240570909178165665757608500558774338041350112205 = 5 x 353 x 1181 x 35401 x 75521 x 160481 x 737501 x 2710260697 x 112096925062539906608469121

296: 324232475273515447634024717929825388762330525546971281881 28597 = 3 x 7 x 73 x 149 x 2221 x 11987 x 10661921 x 54018521 x 81143477963 x 114087288048701953998401

297: 524619165249057853343116499586482964847336113290351695382 40802 = 2 x 17 x 53 x 89 x 109 x 197 x 593 x 4157 x 19801 x1360418597 x 18546805133 x 123692430687502422800 33

298: 84885164052257330097714121751630835360966663883732297726369399 = 110557 x 162709 x 952111 x 4000949 x 4434539 x 85607646594577 x 3263039535803245519

299: 1373470805771631154320257717102791318457002752127674672646 10201 = 233 x 28657 x205699287723427520846348534202713928 20560402848605171521

300: **2222322446294204455297398934619099672066669 39 096499764990979600 = 2^4 x 3^2 x 5^2 x 11 x 31 x 41 x 61 x 101 x 151 x 401 x 601 x 2521 x 3001 x 12301 x 18451 x 570601 x 230686501 x 87129547172401.**
2+2+2+2+3+2+2+4+4+6+2+9+4+2+0+4+4+5+5+2+9+7+3+9+ 8+9+3+4+6+1+9+0+9+9+6+7+2+0+6+6+6+9+3+9+0+9+6+ 4+9+9+7+6+4+9+9+0+9+7+9+6+0+0=315=9

This sequence shows that every 12th number that has a digital root of nine is in a sequence of Tesler's 3, 6 and 9!

12 (**3**) – 24 (**6**) – 36 (**9**) – 48 (12=**3**) – 60 (**6**) – 72 (**9**) – 84 (12 = **3**) – 96 (15 = **6**) – 108 (**9**) – 120 (**3**) – 132 (**6**) – 144 (**9**) – 156 (12 = **3**) – 168 (15 = **6**) – 180 (**9**) – 192 (12 = **3**) – 204 (**6**) – 216 (**9**) – 228 (12 = **3**) – 240 (**6**) – 252 (**9**) – 264 (**12**) – 276 (15 = **6**) – 288 (18 = **9**) – 300 (**3**).

The Simple Complexity of Number Nine

The Fibonacci Numbers that have a digital root of nine (DR9)

144
46368
14930352
4807526976
1548008755920
498454011879264
160500643816367088
51680708854858323072
16641027750620563662096
5358359254990966640871840
1725375039079340637797070384
555565404224292694404015791808
178890334785183168257455287891792
57602132235424755886206198685365216
18547707689471986212190138521399707760
5972304273877744135569338397692020533504
1923063428480944139667114773918309212080528
619220451666590135228675387863297874269396512
199387062373213542599493807777207997205533596336
64202014863723094126901777428873111802307548623680
20672849399056463095319772838289364792345825123228624
6656593304481317393598839952151746590023553382130993248
2143402371193585144275731144820024112622791843221056597232
690168906931029935139391829792095612517948949963798093315456
222232244629420445529739893461909967206666939096499764990979600

Observations:
- The last numbers on the right, going downwards, follow a sequence of 48260!
- The first numbers on the left, going downwards, change their sequence:

14 14 14
15 15 15 15 15
16 16
26 26

The whole tree:
- Left side pattern: 141414 – 1515151515 – 1616 – 2626… NB 1+4=**5** 1+5=**6** 1+6=**7** 2+6=**8**, i.e. running in ascending sequence 5 6 7 8.
- Right side pattern: 48260 48260 48260 48260 48260
- Repeated numbers, from top to bottom: 44 – 66 - 33 – 77 66 – 55 88 00 – 44 44 88 11 – 11 66 66 00 00 33 888 – 555 77 8888 000 33 – 11 666666 0000 222 – 5555 33 888 9999 444 000 666 – 7777777 33333 999 0000 44 – 555555 66 44444 000 2222 99 88 – 1111 77777 8888888 999 333 22 – 555555 77 666666 00 222222 111 44 8888 – 111111 8888 55 44 7777777 0000 666 99999 00 33 –

Repeated Numbers per line in order

44
66
33
66 77
00 55 88
11 4444 88
0000 11 33 6666 888
00 33 555 77 8888
0000 11 222 666666
000 33 444 5555 666 888 9999
0000 33333 44 7777777 999
000 2222 44444 555555 66 88 99
1111 22 333 77777 8888888 999
00 111 222222 44 555555 666666 77 8888
000000 111111 33 44 55 6667777777 8888 99999

Excluding Zero:

$$44$$
$$66$$
$$33$$
$$66\ 77$$
$$55\ 88$$
$$11\ 4444\ 88$$
$$11\ 33\ 6666\ 888$$
$$33\ 555\ 77\ 8888$$
$$11\ 222\ 666666$$
$$33\ 444\ 5555\ 666\ 888\ 9999$$
$$33333\ 44\ 7777777\ 999$$
$$2222\ 44444\ 555555\ 66\ 88\ 99$$
$$1111\ 22\ 333\ 77777\ 8888888\ 999$$
$$111\ 222222\ 44\ 555555\ 666666\ 77\ 8888$$

Other interesting aspect of Fibonacci Numbers:

Sequence 1, 2, 3, 5, 8, 13, 21, 34, 55, 89, 144, 233, 377, 610, 987, 1597, 2584…

The addition of any three consecutive numbers then dividing the sum by two, the answer will always be the third number.

1+2+3 = 6. 6/2 = 3
5+8+13 = 26. 26/2 = 13
8+13+21 = 42. 42/2 = 21
21+34+55 = 110. 110/2 = 55
89+144+233 = 466. 466/2 = 233
377+610+987 = 1974. 1974/2 = 987
1597+2584+4181 = 9362. 8362/2 = 4181

(2) - Lucas Numbers

François Édouard Anatole Lucas (4 April 1842 – 3 October 1891) was a French mathematician who is known for his study of the Fibonacci sequence.

He worked in the Paris observatory and later became a professor of mathematics in Paris.

Lucas was also interested in recreational mathematics. He found an elegant binary solution to the Baguenaudier puzzle. He more notably invented the Tower of Hanoi puzzle, which he marketed under the nickname *N. Claus de Siam*, an anagram of *Lucas d'Amiens*, and published for the first time a description of the Dots and Boxes game in 1889.

Lucas died in unusual circumstances. At the banquet of the annual congress of the *Association française pour l'avancement des sciences*, a waiter dropped some crockery and a piece of broken plate cut Lucas on the cheek. He died a few days later of a severe skin inflammation probably caused by septicaemia. He was only 49 years old.

The Lucas numbers are very similar to the Fibonacci numbers, but start with 2 and 1 (in this order) rather than the Fibonacci's 0 and 1.

n : L_n = Factors of L_n
0: 2
1: 1
2: 3
3: 4 = 2^2
4: 7
5: 11
6: 18 = 2 x 3^2 (1+8=9)
7: 29
8: 47
9: 76 = 2^2 x 19
10: 123 = 3 x 41
11: 199
12: 322 = 2 x 7 x 23
13: 521
14: 843 = 3 x 281
15: 1364 = 2^2 x 11 x 31
16: 2207
17: 3571
18: 5778 = 2 x 3^3 x 107 (5+7+7+8=27=9)

The Simple Complexity of Number Nine

19: 9349
20: 15127 = 7 x 2161
21: 24476 = 2^2 x 29 x 211
22: 39603 = 3 x 43 x 307
23: 64079 = 139 x 461
24: 103682 = 2 x 47 x 1103
25: 167761 = 11 x 101 x 151
26: 271443 = 3 x 90481
27: 439204 = 2^2 x 19 x 5779
28: 710647 = 7^2 x 14503
29: 1149851 = 59 x 19489
30: 1860498 = 2 x 3^2 x 41 x 2521
   (1+8+6+0+4+9+8 = 36 = 9)
31: 3010349
32: 4870847 = 1087 x 4481
33: 7881196 = 2^2 x 199 x 9901
34: 12752043 = 3 x 67 x 63443
35: 20633239 = 11 x 29 x 71 x 911
36: 33385282 = 2 x 7 x 23 x 103681
37: 54018521
38: 87403803 = 3 x 29134601
39: 141422324 = 2^2 x 79 x 521 x 859
40: 228826127 = 47 x 1601 x 3041
41: 370248451
42: 599074578 = 2 x 3^2 x 83 x 281 x 1427
   (5+9+9+7+4+5+7+8 = 54 = 9)
43: 969323029 = 6709 x 144481
44: 1568397607 = 7 x 263 x 881 x 967
45: 2537720636 = 2^2 x 11 x 19 x 31 x 181 x 541
46: 4106118243 = 3 x 4969 x 275449
47: 6643838879
48: 10749957122 = 2 x 769 x 2207 x 3167
49: 17393796001 = 29 x 599786069
50: 28143753123 = 3 x 41 x 401 x 570601
51: 45537549124 = 2^2 x 919 x 3469 x 3571
52: 73681302247 = 7 x 103 x 102193207
53: 119218851371

54: 192900153618 = 2 x 3⁴ x 107 x 11128427
   (1+9+2+9+1+5+3+6+1+8 = 45 = 9)
55: 312119004989 = 11² x 199 x 331 x 39161
56: 505019158607 = 47 x 10745088481
57: 817138163596 = 2² x 229 x 9349 x 95419
58: 1322157322203 = 3 x 347 x 1270083883
59: 2139295485799 = 709 x 8969 x 336419
60: 3461452808002 = 2 x 7 x 23 x 241 x 2161 x 20641
61: 5600748293801
62: 9062201101803 = 3 x 3020733700601
63: 14662949395604 = 2² x 19 x 29 x 211 x 1009 x 31249
64: 23725150497407 = 127 x 186812208641
65: 38388099893011 = 11 x 131 x 521 x 2081 x 24571
66: 62113250390418 = 2 x 3² x 43 x 307 x 261399601
   (6+2+1+1+3+2+5+3+9+4+1+8 = 45 = 9)
67: 100501350283429 = 4021 x 24994118449
68: 162614600673847 = 7 x 23230657239121
69: 263115950957276 = 2² x139 x 461x691x1485571
70: 425730551631123 = 3 x 41 x 281 x 12317523121
71: 688846502588399
72: 1114577054219522 = 2 x47x1103 x10749957121
73: 1803423556807921 = 151549 x 11899937029
74: 2918000611027443 = 3 x 11987 x 81143477963
75: 4721424167835364 = 2² x1 x31x101x151
   x 12301 x 18451
76: 7639424778862807 = 7 x 1091346396980401
77: 12360848946698171 = 29 x 199
   x 229769 x 9321929
78: 20000273725560978 = 2 x 3² x 90481
   x 12280217041
   (2+2+7+3+7+2+5+5+6+9+7+8 = 63 = 9)
79: 32361122672259149
80: 52361396397820127 = 2207 x 23725145626561
81: 84722519070079276 = 2² x 19 x 3079 x 5779
   x 62650261
82: 137083915467899403 = 3 x163 x 800483
   x 350207569
83: 221806434537978679 = 35761381 x 6202401259

The Simple Complexity of Number Nine

84: 358890350005878082 = 2 x 7^2 x 23 x 167 x 14503
x 65740583
85: 580696784543856761 = 11 x 3571
x 1158551 x 12760031
86 : 939587134549734843 = 3 x 313195711516578281
87 : 1520283919093591604 = 2^2 x 59 x 349 x 19489
x 947104099
88: 2459871053643326447 = 47 x 93058241
x 562418561
89: 3980154972736918051 = 179 x
22235502640988369
**90: 6440026026380244498 = 2 x 3^3 x 41 x 107 x 2521
x 10783342081
(6+4+4+2+6+2+6+3+8+2+4+4+4+9+8= 72 = 9)**
91: 10420180999117162549 = 29 x 521
x 689667151970161
92: 16860207025497407047 = 7
x 253367 x 9506372193863
93: 27280388024614569596 = 2^2 x 63799 x 3010349
x 35510749
94: 44140595050111976643 = 3
x 563 x 5641 x 4632894751907
95: 71420983074726546239 = 11 x 191 x 9349
x 41611 x 87382901
96: 115561578124838522882 = 2 x 1087 x 4481
x 11862575248703
97: 186982561199565069121
= 3299 x 56678557502141579
98: 302544139324403592003 = 3 x 281
x 5881 x 61025309469041
99: 489526700523968661124 = 2^2 x 19 x 199
x 991 x 2179 x 9901 x 1513909
100: 792070839848372253127 = 7 x 2161
x 9125201 x 5738108801
101: 1281597540372340914251
= 809 x 7879 x 201062946718741
102: 2073668380220713167378 = 2 x 3^2 x 67 x 409 x

117

63443 x 66265118449
(2+7+3+6+6+8+3+8+2 +7+1+3+1+6+7+3+7+8
= 153 = 9)

103: 3355265920593054081629
= 619 x 1031 x 5257480026438961

104: 5428934300813767249007 = 47
x 3329 x 106513889 x 325759201

105: 8784200221406821330636 = 2^2 x 11 x 29 x 31 x
71 x 211 x 911 x 21211 x 767131

106: 14213134522220588579643 = 3
x 1483 x 2969 x 1076012367720403

107: 22997334743627409910279
= 47927441 x 479836483312919

108: 37210469265847998489922 = 2 x 7 x 23 x 6263 x
103681 x 177962167367

109: 60207804009475408400201
= 128621 x 788071 x 593985111211

110: 97418273275323406890123 = 3 x 41 x 43 x 307
x 59996854928656801

111: 157626077284798815290324 =
2^2 x 4441 x 146521 x 1121101 x 54018521

112: 255044350560122222180447 = 223 x 449 x 2207
x 1154149773784223

113: 412670427844921037470771

114: 667714778405043259651218 = 2 x
3^2 x 227 x 26449 x 29134601 x 212067587
(6+6+7+7+1+4+7+7+8+4+5+4+3+2+5+9+
6+5+1+2+1+8= 108 = 9)

115: 1080385206249964297121989 = 11 x 139 x 461
x 1151 x 5981 x 324301 x 686551

116: 1748099984655007556773207 = 7
x 299281 x 834428410879506721

117: 2828485190904971853895196 = 2^2 x 19 x 79 x
521 x 859 x 1052645985555841

118: 4576585175559979410668403 = 3
x 15247723 x 100049587197598387

119: 7405070366464951264563599 = 29 x 239 x 3571

The Simple Complexity of Number Nine

x 10711 x 27932732439809

120: 11981655542024930675232002 = 2 x 47 x 1103 x 1601 x 3041 x 23735900452321

121: 19386725908489881939795601 = 199 x 97420733208491869044199

122: 31368381450514812615027603 = 3 x 19763 x 21291929 x 24848660119363

123: 50755107359004694554823204 = 2^2 x 4767481 x 370248451 x 7188487771

124: 82123488809519507169850807 = 7 x 743 x 467729 x 33758740830460183

125: 132878596168524201724674011 = 11 x 101 x 151 x 251 x 112128001 x 28143378001

126: 215002084978043708894524818 = 2 x 3^3 x 83 x 107 x 281 x 1427 x 1461601 x 764940961 (2+1+5+2+8+4+9+7+8+4+3+7+8+8+9+4+ 5+2+4+8+1+8 = 117 = 9)

127: 347880681146567910619198829 = 509 x 5081 x 487681 x 13822681 x 19954241

128: 562882766124611619513723647 = 119809 x 4698167634523379875583

129: 910763447271179530132922476 = 2^2 x 6709 x 144481 x 308311 x 761882591401

130: 1473646213395791149646646123 = 3 x 41 x 3121 x 90481 x 42426476041450801

131: 2384409660666970679779568599 = 1049 x 414988698461 x 5477332620091

132: 3858055874062761829426214722 = 2 x 7 x 23 x 263 x 881 x 967 x 5281 x 66529 x 152204449

133: 6242465534729732509205783321 = 29 x 9349 x 10694421739 x 2152958650459

134: 10100521408792494338631998043 = 3 x 6163 x 201912469249 x 2705622682163

135: 16342986943522226847837781364 = 2^2 x 11 x 19 x 31 x 181 x 271 x 541 x 811 x 5779 x 42391 x 119611

136: 26443508352314721186469779407 = 47

x 5626278372832919401376548811

137: 42786495295836948034307560771
= 541721291 x 78982487870939058281

**138: 69230003648151669220777340178 = 2 x 3^2 x 4969 x 16561 x 162563 x 275449 x 1043766587
(6+9+2+3+3+6+4+8+1+5+1+6+6+9+2+2+7+7+7+3+4+1+7+8 = 117 = 9)**

139: 112016498943988617255084900949
= 30859 x 253279129 x 14331800109223159

140: 181246502592140286475862241127 = 7^2 x 2161 x 14503 x 118021448662479038881

141: 293263001536128903730947142076 =
2^2 x 79099591 x 6643838879 x 139509555271

142: 474509504128269190206809383203 = 3 x 283 x 569 x 2820403 x 9799987 x 35537616083

143: 767772505664398093937756525279 = 199 x 521 x 1957099 x 2120119 x 1784714380021

144: 1242282009792667284144565908482 = 2 x 769 x 2207 x 3167 x 115561578124838522881

145: 2010054515457065378082322433761 = 11 x 59 x 19489 x 120196353941 x 1322154751061

146: 3252336525249732662226888342243 = 3 x 29201 x 37125857850184727260788881

147: 5262391040706798040309210776004 = 2^2 x 29 x 211 x 65269 x 620929 x 8844991 x 599786069

148: 8514727565956530702536099118247 = 7 x 10661921 x 114087288048701953998401

149: 13777118606663328742845309894251
= 952111 x 4434539 x 3263039535803245519

**150: 22291846172619859445381409012498 = 2 x 3^2 x 41 x 401 x 601 x 2521 x 570601 x 87129547172401
(2+2+2+9+1+8+4+6+1+7+2+6+1+9+8+5+9+4+4+5+3+8+1+4+9+1+2+4+9+8 = 144 = 9)**

151: 36068964779283188188226718906749 = 1511 x 109734721 x 217533000184835774779

152: 58360810951903047633608127919247 = 47 x 562766385967 x 2206456200865197103

153: 94429775731186235821834846825996 = 2^2 x 19 x 919 x 3469 x 3571 x 13159 x 8293976826829399

154: 15279058668308928345544297474524 3 = 3 x 43 x 281 x 307 x 15252467 x 900164950225760603

155: 24722036241427551927727782157123 9 = 11 x 311 x 3010349 x 29138888651 x 823837075741

156: 40001094909736480273272079631648 2 = 2 x 7 x 23 x 103 x 1249 x 102193207 x 94491842183551489

157: 64723131151164032200999861788772 1 = 39980051 x 1618885 6575286517818849171

158: 10472422606090051247427194142042 03 = 3 x 21803 x 5924683 x 14629892449 x 184715524801

159: 16944735721206454467527180320919 24 = 2^2 x 785461 x 119218851371 x 4523819299182451

160: 27417158327296505714954374462961 27 = 641 x 1087 x 4481 x 8781322404439748742 01601

161: 44361894048502960182481554783880 51 = 29 x 139 x 461 x 1289 x 1917511 x 965840862268529759

162: 71779052375799465897435929246841 78 = 2 x 3^5 x 107 x 11128427 x 1828620361 x 6782976947987

163: 11614094642430242607991748403072 229 = 1043201 x 6601501 x 1686454671192230445929

164: 18791999880010189197735341327756 407 = 7 x 268457141143 0027028247905903965201

165: 30406094522440431805727089730828 636 = 2^2 x 11^2 x 31 x 199 x 331 x 9901 x 39161 x 51164521 x 1550853481

166: 49198094402450621003462431058585 043 = 3 x 6464041 x 245329617161 x 10341247759646081

167: 79604188924891052809189520789413 679 = 766531 x 103849927 6935845423201273 27909

168: 12880228332734167381265195184799 8722 = 2 x 47 x 1103 x 10745088481 x 115613939510481515041

169: 20840647225223272662184147263741 2401 = 521 x 596107814364089 x 671040394220849329

170: 33720875557957440043449342448541 1123 = 3 x 41 x 67 x 1361 x 40801 x 63443 x 11614654211954032961

171: 5456152278318071270563348971228235249 = 2^2 × 19^2 × 229 × 9349 × 95419 × 162451 × 1617661 × 7038398989

172: 882823983411381527490828321608234647 = 7 × 1261177119159 × 11646784404045944033521

173: 1428439211243188654547163218731058171 = 78889 × 6248069 × 16923049609 × 171246170261359

174: 2311263194654570182037991540339292818 = 2 × 3^2 × 347 × 97787 × 528295667 × 1270083883 × 5639710969

175: 3739702405897758836585154759070350989 = 11 × 29 × 71 × 101 × 151 × 911 × 54601 × 560701 × 7517651 × 51636551

176: 6050965600552329018623146299409643807 = 1409 × 2207 × 6086461133983 × 319702847642258783

177: 9790668006450087855208301058479994796 = 2^2 × 709 × 8969 × 336419 × 10884439 × 105117617351706859

178: 15841633607002416873831447357889638603 = 3 × 5280544535 × 667472291277149119296546201

179: 25632301613452504729039748416369633399 = 359 × 106673 × 7847220321 × 66932254279484647441

180: 41473935220454921602871195774259272002 = 2 × 7 × 23 × 241 × 2161 × 8641 × 20641 × 103681 × 13373763765986881

181: 67106236833907426331910944190628905401 = 97379 × 213732 × 61504197751 × 32242356485644069

182: 108580172054362347934782139964888177403 = 3 × 281 × 90481 × 232961 × 6110578634294886534808481

183: 175686408888269774266693084155517082804 = 2^2 × 14686239709 × 5600748293801 × 533975715909289

184: 284266580942632122201475224120405260207 = 47 × 367 × 37309023160481 × 441720958100381917103

185: 459952989830901896468168308275922343011 = 11 × 54018521 × 265272771839851 × 2918000731816531

186: 744219570773534018669643532396327603218 = 2 × 3^2 × 15917507 × 3020733700601 × 859886421593527043

187: 1204172560604435915137811840672249946229 = 199 × 1871 × 3571 × 905674234408506526265097390431

188: 1948392131377969933807455373068577549447 = 7 x 18049 x 100769 x 153037630649666194962091443041

189: 3152564691982405848945267213740827495676 = 2² x 19 x 29 x 211 x 379 x 1009 x 5779 x 31249 x 85429 x 912871 x 1258740001

190: 5100956823360375782752722586809405045123 = 3 x 41 x 2281 x 4561 x 29134601 x 782747561 x 174795553490801

191: 8253521515342781631697989800550232540799 = 22921 x 395586472506832921 x 910257559954057439

192: 13354478338703157414450712387359637585922 = 2 x 127 x 383 x 5662847 x 6803327 x 19073614849 x 186812208641

193: 21607999854045939046148702187909870126721 = 303011 x 76225351 x 935527893146187207403151261

194: 34962478192749096460599414575269507712643 = 3 x 195163 x 4501963 x 5644065667 x 2350117027000544947

195: 56570478046795035506748116763179377839364 = 2² x 11 x 31 x 79 x 131 x 521 x 859 x 1951 x 2081 x 2731 x 24571 x 866581 x 37928281

196: 91532956239544131967347531338448885552007 = 7³ x 14503 x 3016049 x 6100804791163473872231629367

197: 148103434286339167474095648101628263391371 = 31498587119111339 x 4701907222895068350249889

198: 239636390525883299441443179440077148943378
= 2 x 3³ x 43 x 107 x 307 x 261399601
x 11166702227 x 1076312899454363

199: 387739824812222466915538827541705412334749 = 2389 x 4503769 x 36036960414811969810787847118289

200: 627376215338105766356982006981782561278127 = 47 x 1601 x 3041 x 124001 x 6996001 x 3160438834174817356001

Note the similarity with Fibonacci numbers – every 12th number has a digital sum of 9!

(3) - The Decimal Expansion of the Golden Ratio

Take the ratio of two successive numbers in Fibonacci's series, divide each by the number before it, you will find the following series of numbers:

1/1 = 1, 2/1 = 2, 3/2 = 1.5, 5/3 = 1.666..., 8/5 = 1.6, 13/8 = 1.625, 21/13 = 1.61538...

If you plot a graph of these values you'll see that they seem to be tending to a limit, which we call the *golden ratio* (also known as the *golden number* and *golden section*).

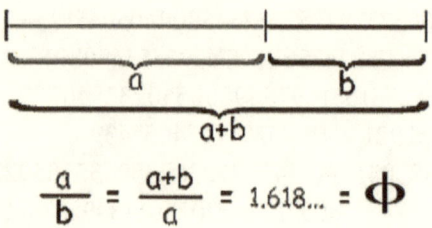

$$\frac{a}{b} = \frac{a+b}{a} = 1.618... = \Phi$$

Phi (Φ) = 1.61803398874989484204586834365...

<u>1, 6, 1, 8, 0, 3, 3, 9, 8, 8, 7</u>, <u>4, 9, 8, 9, 4, 8, 4, 8</u>, <u>2, 0, 4, 5, 8, 6, 8, 3</u>, <u>4, 3, 6, 5</u>, <u>6, 3, 8, 1</u>, <u>1, 7, 7, 2, 0, 3, 0, 9, 1, 7, 9, 8</u>, <u>0, 5, 7, 6</u>, <u>2, 8, 6, 2</u>, <u>1, 3, 5, 4, 4, 8, 6, 2, 2, 7, 0, 5, 2, 6, 0, 4, 6, 2, 8, 1, 8, 9, 0, 2, 4, 4, 9, 7</u>, <u>0, 7, 2</u>, <u>0, 7, 2</u>, <u>0, 4, 1, 8, 9, 3, 9, 1, 1</u>, 3, 7, 4, 8, 4, 7, 5...

1, 6, 1, 8, 0, 3, 3, 9, 8, 8, 7, (=54=9) 4, 9, 8, 9, 4, 8, 4, 8, (=54= 9) 2, 0, 4, 5, 8, 6, 8, 3, (=36=9) 4, 3, 6, 5, (=18=9) 6, 3, 8, 1, (=18=9) 1, 7, 7, 2, 0, 3, 0, 9, 1, 7, 9, 8, (=54=9) 0, 5, 7, 6, (=18=9) 2, 8, 6, 2, (=18=9) 1, 3, 5, (=9) 4, 4, 8, 6, 2, 2, 7, 0, 5, 2, 6, 0, 4, 6, 2, 8, 1, 8, 9, 0, 2, 4, 4, 9, 7, (=117=9) 0, 7, 2, (=9) 0, 7, 2, (=9) 0, 4, 1, 8, 9, 3, 9, 1, 1, (=36=9) 3, 7, 4, 8, 4, 7, 5...

(4) - Prime Numbers

Table of prime numbers up to 1,000:

<u>2, 3, 5, 7, 11, 13, 17, 19, 23, 29, 31, 37, 41, 43, 47, 53, 59, 61, 67, 71</u> **(639=18=9.** N=20), <u>73, 79, 83, 89</u> **(324=9.** N=4) <u>97, 101</u> **(198=18=9.** N=2) <u>103, 107, 109, 113</u> **(432=9.** N=4), <u>127, 131, 137, 139,149, 151, 157, 163, 167, 173</u> **(1494=18=9.** N=10), <u>179,181</u> **(360=9.** N=2), <u>191,193,197, 199, 211, 223, 227, 229, 233, 239, 241, 251, 257, 263, 269, 271, 277, 281, 283, 293, 307, 311, 313, 317, 331, 337</u> **(4023=9.** N=26)) <u>347,349, 353, 359, 367, 373, 379, 383, 389, 397, 401, 409, 419, 421</u> **(5346=18=9.** N=14), <u>431, 433</u> **(864=18=9.** N=2) <u>439, 443</u>, **(882=18=9. N=2)** <u>449, 457, 461, 463, 467, 479, 487, 491, 499, 503, 509</u>**(5265=18=9.** N=11), <u>521, 523</u> **(1044=9 N=2),**

THE SIMPLE COMPLEXITY OF NUMBER NINE

<u>541, 547, 557, 563, 569, 571, 577, 587, 593, 599, 601, 607</u> (**6912=18=9. N=12**), 613, 617, 619, 631, 641, 643, 647, 653, 659, 661, 673, 677, 683, 691, 701, 709, 719, 727, 733, 739, 743, 751, 757, 761, 769, 773, 787, 797, 809, 811, 821, 823, 827, 829, 839, 853, 857, 859, 863, 877, 881, 883, 887, 907, 911, 919, 929, 937, 941, 947, 953, 967, 971, 977, 983, 991, 997

The first 9 numbers DS = 100 DR = 1

The first twenty numbers:
- Number of digits = 36 = 9
- DS = 639
- DR = 9 (6+3+9=18=9)

The proceeding four numbers:
- DS = 324
- DR = 9

The following two numbers:
- DS = 198
- DR = 9

The following four numbers:
- DS = 432
- DR = 9

The following ten numbers:
- DS = 1494
- DR = 9

The following two numbers:
- DS = 360
- DR = 9

The following twenty six numbers:
- DS = 4023
- DR = 9

The following fourteen numbers:
- DS = 5346
- DR = 9

The following two numbers:
- DS = 864
- DR = 9

The following two numbers:
- DS = 882
- DR = 9

The following eleven numbers:
- DS = 5265
- DR = 9

The following two numbers:
- DS = 1044
- DR = 9

The following twelve numbers:
- DS = 6912
- DR = 9

The number of digits which sum up to nine, as above equals to 108: (9+20+4+2+4+10+2+26+14+2+2+11+2 = 108) 1+8 = 9

(5) - Decimal Expansion of the Plastic Constant

The plastic constant or plastic number **P**, also called the silver constant or silver number, is the real root of the cubic equation '$x^3 - x - 1 = 0$'.

P =
1.3247179572447460259609088544780 9734…

1, 3, 2, 4, 7, 1, 7, 9, 5, 7, 2, 4, 4, 7, 4, 6, 0, 2, 5, 9, 6, 0, 9, 0, 8, 8, 5, 4, 4, 7, 8, 0, 9, 7, 3, 4, 0, 7, 3, 4, 4, 0, 4, 0, 5, 6, 9, 0, 1, 7, 3, 3, 3, 6, 4, 5, 3, 4,…

The first six numbers:
- DS = 18
- DR = 9

The following eight numbers:
- DS = 45
- DR = 9

The following twenty two numbers:
- DS = 108
- DR = 9

The following nine numbers:
- DS = 27
- DR = 9

The following thirteen numbers:
- DS = 54
- DR = 9

(6) - Cube Numbers (k)

A cube number is the product of three equal factors of natural numbers.
$k = n \times n \times n = n^3$
Example: 2 x 2 x 2 = 8. 8 is a cube number.
  9 x 9 x 9 = 729. 729 is a cube number.

The first 100 cube numbers:
1   8   27   64   125   216   343   512   729   1000
1331   1728   2137   2744   3375   4096   4913
5832   6859   8000   9261   10648   12167   13824
15625   17576   19683   21952   24389   27000
29791   32768   35937   39304   42875   46656
50653   54872   59319   64000   68921   74088
79507   85184   91125   97336   103823   110592

117649 125000 <u>132651</u> 140608 148877
<u>157464</u> 166375 175616 <u>185193</u> 195112
205379 <u>216000</u> 226981 238328 <u>250047</u>
262144 274625 <u>287496</u> 300763 314432
<u>328509</u> 343000 357911 <u>373248</u> 389017
405224 <u>421875</u> 438976 456533 <u>474552</u>
493039 512000 <u>531441</u> 551368 571787
<u>592704</u> 614125 636056 <u>658503</u> 681472
704969 <u>729000</u> 753571 778688 <u>804357</u>
830584 857375 <u>884736</u> 912673 941192
<u>970299</u> 1000000

27: 2+7=9.
216: 2+1+6=9.
729: 7+2+9=18=9.
1728: 1+7+2+8=18=9.
3375: 3+3+7+5=18=9.
5832: 5+8+3+2=18=9.
9261: 9+2+6+1=18=9.
13824: 1+3+8+2+4=18=9.
19683: 1+9+6+8+3=27=9.
27000= 2+7=9.
35937: 3+5+9+3+7=27=9.
46656: 4+6+6+5+6=27=9.
59319: 5+9+3+1+9=27=9.
74088: 7+4+8+8=27=9.
91125: 9+1+1+2+5=18=9.
110592: 1+1+5+9+2=18=9.
132651: 1+3+2+6+5+1=18=9.
157464: 1+5+7+4+6+4=27=9.
185193: 1+8+5+1+9+3=27=9.
216000: 2+1+6=9.
250047: 2+5+4+7=18=9.
287496: 2+8+7+4+9+6=36=9.
328509: 3+2+8+5+9=27=9.
373248: 3+7+3+2+4+8=27=9.

421875: 4+2+1+8+7+5=27=9.
474552: 4+7+4+5+5+2=27=9.
531441: 5+3+1+4+4+1=18=9.
592704: 5+9+2+7+4=27=9.
658503: 6+5+8+5+3=27=9.
729000: 7+2+9=18=9.
804357: 8+4+3+5+7=27=9.
884736: 8+8+4+7+3+6=36=9.
970299: 9+7+2+9+9=36=9.

Every third cube number is a DR9!

Since a cube number is n x n x n (n^3), then every third cube number should have a digital sum or digital root equal that equal digital root of 9.

Further analysis of cube numbers:

	1	2	3	4	5	6	7	8	9	0
	1	8	2	6	12	21	34	51	72	10
			7	4	5	6	3	2	9	00
1	13	17	21	27	33	40	49	58	68	80
	3	2	9	4	7	9	1	3	5	0
	1	8	7	4	5	6	3	2	9	0
2	92	10	12	13	15	17	19	21	24	27
	6	64	16	82	62	57	68	95	38	00
	1	8	7	4	5	6	3	2	9	0
3	29	32	35	39	42	46	50	54	59	64
	79	76	93	30	87	65	65	87	31	00
	1	8	7	4	5	6	3	2	9	0
4	68	74	79	85	91	97	10	11	11	12
	9	0	5	1	1	3	38	05	76	50
	2	8	0	8	2	3	2	9	4	0
	1	8	7	4	5	6	3	2	9	0
5	13	14	14	15	16	17	18	19	20	21
	26	06	88	74	63	56	51	51	53	60
	5	0	7	6	7	1	9	1	7	0
	1	8	7	4	5	6	3	2	9	0
6	22	23	25	26	27	28	30	31	32	34
	69	83	00	21	46	74	07	44	85	30
	8	2	4	4	2	9	6	3	0	0
	1	8	7	4	5	6	3	2	9	0
7	35	37	38	40	42	43	45	47	49	51
	79	32	90	52	18	89	65	45	30	20
	1	4	1	2	7	7	3	5	3	0
	1	8	7	4	5	6	3	2	9	0
8	53	55	57	59	61	63	65	68	70	72
	14	13	17	27	41	60	85	14	49	90
	4	6	8	0	2	5	0	7	6	0
	1	8	7	4	5	6	3	2	9	0
9	75	77	80	83	85	88	91	94	97	10
	35	86	43	05	73	47	26	11	02	00
	7	8	5	8	7	3	7	9	9	00
	1	8	7	4	5	6	3	2	9	0

The Simple Complexity of Number Nine

All numbers in every column end with the same number:

Column 1 ends with 1, column 2 ends with 8, column 3 ends with 7, column 4 ends with 4, column 5 ends with 5, column 6 ends with 6, column 7 ends with 3, column 8 ends with 2, column 9 ends with 9, and column 10 ends with 0!

Examining the rows also reveals interesting similarities:

Column 1 – Numbers in rows 5 to 9 start with 1,2,3,5 & 7.
Column 2 – Numbers in rows 5 to 9 start with 1,2,3,5 & 7!
Column 3 – Numbers in rows 5 to 9 start with 1,2,3,5 & 8.
Column 4 – Numbers in rows 5 to 9 start with 1,2,3,5 & 8!
Column 5 – Numbers in rows 5 to 9 start with 1,2,4,6 & 8.
Column 6 – Numbers in rows 5 to 9 start with 1,2,4,6 & 8!
Column 7 – Numbers in rows 5 to 9 start with 1,3,4,6 & 9.
Column 8 – Numbers in rows 5 to 9 start with 1,3,4,6 & 9!

Other interesting observations:

Column 4 – the numbers before the last number run downwards in a pattern, as such: 64208 & 64208!

Column 5 – the numbers before the last number run downwards in a pattern, as such: 27, 27, 27, 27 & 27!

Column 5 – the numbers before two numbers from the end run downwards in a pattern, as such: 1368& 1368 (13...)

Column 6 – the numbers before the last number run downwards in a pattern, as such: 19753 & 19753.

Column 8 – the numbers before the last number run downwards in a pattern, as such: 13579 & 13579.

Nine Around the World

- Cultures and Religions
  1-Aztecs
  2-Babylonian
  3-Buddhism
  4-Brahmanism
  5-Celtic
  6-Chinese
  7-Druidism
  8-Egyptian
  9-European
  10-Greek mythology
  11-Hebrew
  12-Hindu
  13-Islam
  14-Japanese
  15-Mayan
  16-Nepalese
  17-Orphism
  18-Pythagorean
  19-Roman
  20-Scandinavian
- Idioms and popular phrases
- Literature
- History
- Mathematics
- Sciences
- Sports
- Languages
- Nine in different languages
- Miscellaneous

Cultures and Religions

Aztecs
- The nine stages that should traverse the souls of Aztecs to reach the eternal rest.
- They counted also nine underground worlds.

Babylon
The Babylonian counting system was based on the number 60, which was almost certainly chosen because it was the lowest number to have 2, 3, 4, 5 and 6 as factors (you need to get to 420 to get 7, and 60 also has as factors 10, 12, 15, 20 and 30)

Baha'i Faith
- Nine, as the highest single-digit number (in base ten), symbolizes completeness in the Baha'í Faith. In addition, the word Baha' in the Abjad notation has a value of 9, and a 9-pointed star is used to symbolize the religion.
- The symbol of the Baha'i Faith is a 9-pointed star.

Buddhism
- Important Buddhist rituals usually involve nine monks.
- In Hindu and Buddhist cultures, the number 108 is a sacred one (1+8=9). There are 108 Buddhist saints. The Nepalese parliament has 108 seats.
- In Buddhism, Gautama Buddha, was believed to have nine virtues, which he was (1) Accomplished, (2) Perfectly Enlightened, (3) Endowed with knowledge and Conduct or Practice, (4) Well-gone or Well-spoken, (5) the Knower of worlds, (6) the Guide Unsurpassed of men to be tamed, (7) the Teacher of gods and men, (8) Enlightened, and (9) Blessed.
- Important Buddhist rituals usually involve nine monks.

Brahmanism
- Vichnou (Vishnu) incarnates in nine avatars to sacrifice himself for the salvation of men.

- One day in the life of Brahma is 4,320,000,000 years (4+3+2=0), and a year of Brahma is 360 of his days and nights (3,110,400,000,000 years – 3+1+1+4=9).
- The lifetime of Brahma is a hundred of these, and after two Brahman lifetimes, the cycle of time starts all over again in an identical repetition of 622,080,000,000,000 years (6+2+2+8= 18 = 9).

Celtic

- Celtic legend symbolizes nine as a highly significant number.
- It is a central number with the eight directions with the centre making nine.
- The Triple Goddesses are thrice three.
- There are nine Celtic maidens and nine white stones that symbolize the nine virgins' attendant on Bridgit.
- Nine is connected with the Beltane Fire rites which are attended by 81 (8+ 1= 9) men, nine at a time.

Chinese

- Nine is a celestial power. It is 3x3 being the most auspicious of all the numbers.
- Nine also signifies the eight directions with the centre as the ninth point known as the Hall of Light.
- There are nine great social laws and classes of officials.
- In land divisions for FengShui there are eight exterior squares for cultivation of the land by holders and the central, and ninth, square is a "god's acre", dedicated to Shang-ti, the supreme ruler. It is also known as the Emperor's Field, giving homage and respect denoting the position of heavenly power.
- Nine (pinyin jiǔ) is considered a good number in Chinese culture because it sounds the same as the word "long-lasting" (pinyinjiǔ).
- Nine is strongly associated with the Chinese dragon, a symbol of magic and power.
- There are nine forms of the dragon, it is described in terms of nine attributes, and it has nine children. It has 117

scales - 81 yang (masculine, heavenly) and 36 yin (feminine, earthly). All three numbers are multiples of 9 (9×13=117, 9×9=81, 9×4=36) as well as having the same digital root of 9.

- The dragon often symbolizes the Emperor, and the number nine can be found in many ornaments in the Forbidden City.
- A Chinese dragon has 117 scales (1+1+7=9). This is because of the significance of number nine in Chinese culture, where it is thought to be lucky. 81 scales (8+1=9) are male and 36 (3+6=9) are female. Note: 9×9 = 81 & 4×9 = 36…
- There are 9 types of dragons in Chinese folklore: Celestial Dragon, Spiritual Dragon, Dragon of Hidden Treasures, Underground Dragon, winged Dragon, Horned Dragon, Coiling Dragon, Yellow Dragon, and Dragon King.
- The circular altar platform (*Earthly Mount*) of the Temple of Heaven has one circular marble plate in the centre, surrounded by a ring of nine plates, then by a ring of 18 plates, and so on, for a total of nine rings, with the outermost having 81=9×9 plates.
- The name of the area called Kowloon in Hong Kong literally means *nine dragons*.
- The (Chinese: 南海九段线; pinyin: *nánhǎijiǔduànxiàn*; literally: "Nine-segment line of the South China Sea") delimits certain island claims by China in the South China Sea.
- The nine-rank system was a civil service nomination system used during certain Chinese dynasties.
- There are 81 chapters (8+1=9) in the ancient Chinese book, Tao TeChing.

Christianity
- Nine is one of the numbers that appears scantly in Christian symbolism.
- There are the triple triads of choirs of angels and nine spheres and nine rings around hell.
- Jesus appears nine times to his disciples and apostles after his resurrection.
- Nailed on the cross, Jesus Christ expires at the ninth hour.
- The nine Choruses of the Angels

- After the death of a Pope, we celebrate masses for the rest of his soul during nine days, with nine absolutions.
- The number 9 is used 50 times in the Bible.
- In the Christian angelic hierarchy, there are 9 choirs of angels.
- The books of the New Testament are 27 (2+7=9): Matthew. Ephesians. Hebrews. Mark. Philippians. James. Luke. Colossians. Peter I. John. Thessalonians I. Peter II. John I. Acts. Thessalonians II. Romans. Timothy I. John II. Corinthians I. Timothy II. John III. Corinthians II. Titus. Jude. Galatians. Philemon. Revelation.
- Jesus is said to have sent out 72 (7+2=9) disciples.

Druidism
The 9 girls of Belenos of the druidism: Ogia - virginity, Glania - purity, Karantia - charity, Uxellia - nobleness, Viriona - truth, Aventia - honesty, Dagia - goodness, Lania - plenitude, and Lovania - joy.

Egyptian
- The nine bows is a term used in Ancient Egypt to represent the traditional enemies of Egypt
- The Ennead is a group of 9 Egyptian deities, who, in the some versions of the Osiris myth, judged whether Horus or Set should inherit Egypt.

European Culture
- The Nine Worthies are nine historical or semi-legendary figures who, in the Middle Ages, were believed to personify the ideals of chivalry

Freemasonry
- In Freemasonry, there is an Order of "Nine Elected Knights," and in the working of this Order 9 roses, 9 lights and 9 knocks must be used.

Greek Mythology
- The nine Muses in Greek mythology are Calliope (epic poetry), Clio (history), Erato (erotic poetry), Euterpe (lyric poetry),

Melpomene (tragedy), Polyhymnia (song), Terpsichore (dance), Thalia (comedy), and Urania (astronomy).
- It takes nine days (for an anvil) to fall from heaven to earth, and nine more to fall from earth to Tartarus - a place of torment in the underworld.
- Leto laboured for nine nights and nine days for Apollo, according to the Homeric Hymn to Delian Apollo.
- The nine nights of love of Zeus.
- The nine days and nine nights, that Leto suffers when she gives birth.
- The nine girls of Zeus and Mnemosyne, named Muses, who governed the liberal arts: Clio, Calliope, Melpomene, Thala, Euterpe, Erato, Terpsichore, Polymna and Urana.
- The nine days of anxiety of the Ceres-Demeter goddess who went around the world in search of her daughter Persephone kidnaped by Hephaistos, the dark god of Hells;
- The tradition wants that Minos, in his cavern, spent nine years to receive the Jupiter laws; a legend says also that Minos had, every nine years, a meeting with Jupiter, after what it was possible for him to prophesy.
- Troy was besieged for **9** days and fell on the tenth.
- Odysseus (Lat. Ulysses) wandered for 9 years and arrived home on the tenth.

Hebrew
- Nine is pure intelligence (eight was perfect intelligence). Also represents truth, since it reproduces itself when multiplied.
- In Cabbalism, nine symbolizes foundation.
- In the Kabbalah the number 216 (2+1+6=9) is believed to represent the name of God.
- The first nine days of the Hebrew month of Av are collectively known as "The Nine Days" (TishaHaYamim), and are a period of semi-mourning leading up to TishaB'Av, the ninth day of Av on which both Temples in Jerusalem were destroyed.

Hindu
- Nine is the number of Agni, fire.
- The square of the nine forms the mandala of eighty-one squares and leads to, and encloses the Universe.
- The number 9 is revered in Hinduism and considered a complete, perfect and divine number. They believe that it represents the end of a cycle in the decimal system, which originated from the Indian subcontinent as early as 3000 BC.
- In Hinduism the number 432 (4+3+2=9) is an important factor in its traditions. The dark age in the Hindu tradition is equal to 432,000 years, during which the morality of humans degenerate.

Islam
- Islam acknowledges the 99 Beautiful Names of God.
- Ramadan, the month of fasting and prayer, is the ninth month of the Islamic calendar.
- Islamic cosmology the universe is made from 9 spheres - the traditional 8 of Ptolemy, plus a ninth added by the Arab astronomer Thabit ibn Qurra to explain the precession of the equinoxes.

Japanese:
The Japanese consider nine to be unlucky because in Japanese the word for nine sounds similar to the word for "pain" or "distress".

Mayan:
There are nine underworlds each ruled by a God. We find this reference to "nine underworlds" present in many cultures and beliefs.

Nepal
There is nine "Dharmas", books constituting the Nepal Bible.

Orphism
Orphism defines nine symbolic aspects of the universe divided into 3 triads: the night, the sky and the time; the ether, the light and the stars; the sun, the moon and the nature.

Pythagorean:
- Pythagoras, and his followers saw the universe as not only harmonious, but also harmonic. For them philosophy, religion, music and mathematics were all aspects of a single, unified idea. This was one of first great formulations of a philosophy of numbers in history.
- The nine is the limit of all numbers, all others existing and coming from the same. i.e.: 0 to 9 is all one needs to make up an infinite amount of numbers.

Roman
- The Romans held a feast in memory of their dead every 9th year.
- Nundinae ("9th day", the Roman market day): From earliest times, a market day occurred every 9th day, according to the Roman method of inclusive reckoning. It was a day of rest from agricultural labour and a time to take produce to market.

Scandinavian:
- Odin/Woden hung for nine days and nights on the Yggdrasil to win the secrets of wisdom for humankind.
- Skeldi, the northern Persephone, the goddess of snow, lives in her mountain for three months and by Niord's sea for nine months.
- Nine is the sacred number in Scandinavian-Teutonic symbolism.
- In Norse mythology, the Doomsday of the gods will feature 800 warriors coming through the 540 doors of Valhalla: a total of 432,000. (5+4=9). (4+3+2=9).

Taoism
"Jiugong" (also known as "Jiuxuan"), in the Mystical Numbers of **Taoism**, represents the number 9 and symbolizes the Nine Chambers. It is the last number of order before returning to chaos. After that, the cycle repeats...

The Sacred Numbers 54, 108 and 216.
(5+4=9   1+0+8=9   2+1+6=9)
The number 108 comes up in a variety of contexts.
- There are 54 letters in the Sanskrit alphabet. Each has a masculine/feminine or Shiva/Shakti aspect: 54 x 2 is 108. On the sacred

geometrical configuration known as the Sri Yantra, there are points called marmas where three lines intersect, and there are 54 such intersections. Each intersection also has masculine/feminine or Shiva/Shakti qualities.

- In the human body, marmas or marmastanas are energy intersections similar to chakras. There are said to be 108 marmas in the subtle body.
- It is also said that there are a total of 108 main energy lines or 'nadis' that radiate from the heart chakra. Thus, the number 108 relates to the Sri Yantra as well as the human body.
- The diameter of the sun is 108 times the diameter of the Earth. In astrology, there are 12 houses and 9 planets.
- In the Krishna tradition, there was said to be 108 gopis or maidservants of Krishna. There are 108 forms of traditional Indian dance. In the Vedas, there are 108 Upanishads as listed in the Mundakopanishad. There is said to be 108 Tantric texts. Ancient texts say that the Atman, the human soul, passes through 108 different stages on its path to enlightenment. The list goes on and on.
- Sikh: The Sikh tradition has a mala of 108 knots tied in a string of wool, rather than beads.
- Buddhism: Some Buddhists carve 108 small Buddhas on a walnut for good luck. Some ring a bell 108 times to celebrate a new year. It is said that there are to be 108 virtues to cultivate and 108 defilements to avoid.
- Chinese: The Chinese Buddhists and Taoists use a 108-bead mala, which is called su-chu, and has three dividing beads, so the mala is divided into three parts of 36 each.
- Chinese astrology says that there are 108 sacred stars. Stages of the soul: It is said that Atman, the human soul or center goes through 108 stages on the journey.
- Nine dishes are traditionally served at a Chinese birthday banquet.
- Number nine is associated with longevity in the Chinese culture.

The 3 by 3 Magic Square

The 3 by 3 square is revered in the cultures of Islam, Jains of India, Tibetan Buddhism, Celts, African, Shamanic, and Jewish mysticism.

```
4 9 2
3 5 7
8 1 6
```

In this arrangement, the three columns, three rows and two diagonals always add up to fifteen. In *Feng Shui* the numbers within each cell of the magic square have specific significance for working with the earth's subtle creative energies for the good of society and the environment.

Idioms and popular phrases

- 'to go the whole nine yards'
- 'A cat has nine lives'
- 'to be on cloud nine'
- 'A stitch in time saves Nine'
- 'The whole nine yards'
- 'found true 9 out of 10 times'
- The word "K-9" pronounces the same as *canine* and is used in many U.S. police departments to denote the police dog unit. Despite not sounding like the translation of the word *canine* in other languages, many police and military units around the world use the same designation.
- Someone dressed "to the nines" is dressed up as much as they can be.
- In urban culture, "nine" is a slang word for a 9 mm pistol or homicide, the latter from the Illinois Criminal Code for homicide.
- 'Possession is nine parts of the law' is an expression used in legal disputes over ownership.
- 'As bent as a nine-bob note' is English slang for something that is not bona fida. 'Bob' was the slang for the old British shilling, which came in ones, twos, and tens; but not in nines.
- 'Dressed to the nine's' means dressed very glamorously. Of unknown origin.

Literature

- There are nine circles of Hell in Dante's Divine Comedy.
- The Nine Bright Shiners, characters in Garth Nix's Old Kingdom trilogy. The Nine Bright Shiners was a 1930s book of poems by Anne Ridderand a 1988 fiction book by Anthea Fraser; the name derives from "a very curious old semi-pagan, semi-Christian" song.
- The Nine Tailors is a 1934 mystery novel by British writer Dorothy L. Sayers, her ninth featuring sleuth Lord Peter Wimsey.
- Nine Unknown Men are, in occult legend, the custodians of the sciences of the world since ancient times
- In J.R.R. Tolkien's Middle-earth, there are nine rings of power given to men, and nine ring wraiths. Additionally, The Fellowship of the Ring consists of nine companions, representing the free races and also as a positive mirror of the nine ring wraiths.
- In Lorien Legacies there are nine Garde sent to Earth.
- Number 9 is a character in Lorien Legacies
- The nine levels of the hell of Dante.

History

Columbus
Sailed on 03.08.1492 = 27 = 9
Landed: 02.10.1492. 2x1x1x4x9x2 = 144.
       1+4+4 = 9
Journey took 37 days.
  73-37 = 36. 3+6 = 9.
Died on 07.12.1506. = 64080 = 18 = 9

60512170 - 07121506 = 53390664
  5+3+3+9+6+6+4 = 36 = 9
6051 - 1506 = 4545. 4+5+4+5 = 18 = 9
712 - 217 = 495. 4+9+5 = 18 = 9

The Simple Complexity of Number Nine

Alexander the Great
Born 356 BC
Founded Alexandria 332 BC
Crowned at Ammon Temple, Siwa 330 BC
Died at Babylon 323 BC, aged 33 years.
356: 3x5x6 = 90 = 9
332 & 323: 3x3x2 = 18 = 9
330: 3x3 = 9
33: 3x3 = 9

Mathematics

Plato's vision of mathematics: In the Republic, a dialogue of Plato from the 4th century BC, the philosopher Socrates and his companion Glaucon explore the nature and value of mathematics as a branch of knowledge. Socrates asks, "What would be the study that would draw the soul away from the world of becoming to the world of being?" They, Socrates and Glaucon, determine that neither gymnastics nor the arts can do this, but only a science that is universal and studies the truth.

- Nine is a composite number, its proper divisors being 1 and 3. It is 3 times 3 and hence the third square number.
- Nine is the highest single-digit number in the decimal system. It is the second non-unitary square prime of the form and the first that is odd. All subsequent squares of this form are odd. It has a unique aliquot sum 4, which is itself a square prime.
- Nine is; and can be, the only square prime with an aliquot sum of the same form. The aliquot sequence of nine has 5 members (9,4,3, 1,0) this number being the second composite member of the 3-aliquot tree. It is the aliquot sum of only one number the discrete semi-prime 15.
- The number 9 is an emblem of matter that can never be destroyed, so the number 9 when multiplied by any number always reproduces itself, no matter what the extent of the number is that has been employed.
- Eight and 9 form a Ruth-Aaron pair under the second definition that counts repeated prime factors as often as they occur.

- A polygon with nine sides is called a nonagon or enneagon. A group of nine of anything is called an ennead.
- Number 999 (9+9+9=27=9) is the first emergency telephone service, which was implemented in 1937 in London.
- The Empire State Express was the first train to exceed 100 mph, and its number was 999!
- It is to be noted that 999 is 666 (6+6+6=18=9) rotated by 180 degrees (1+8=9).
- In base 10, a positive number is divisible by nine if and only if its digital root is 9. That is, if you multiply nine by any natural number, and repeatedly add the digits of the answer until it is just one digit, you will end up with nine:
  - 2 × 9 = 18 (1 + 8 = 9)
  - 3 × 9 = 27 (2 + 7 = 9)
  - 9 × 9 = 81 (8 + 1 = 9)
  - 121 × 9 = 1089 (1 + 0 + 8 + 9 = 18; 1 + 8 = 9)
  - 234 × 9 = 2106 (2 + 1 + 0 + 6 = 9)
  - 578329 × 9 = 5204961 (5 + 2 + 0 + 4 + 9 + 6 + 1 = 27; 2 + 7 = 9)
  - 482729235601 × 9 = 4344563120409 (4 + 3 + 4 + 4 + 5 + 6 + 3 + 1 + 2 + 0 + 4 + 0 + 9 = 45; 4 + 5 = 9)

There are other interesting patterns involving multiples of nine:

- 12345679 x 9 = 111111111
- 12345679 x 18 = 222222222
- 12345679 x 81 = 999999999

This works for all the multiples of 9. n = 3 is the only other n > 1 such that a number is divisible by n if and only if its digital root is n.

The difference between a base-10 positive integer and the sum of its digits is a whole multiple of nine.

Examples:

- The sum of the digits of 41 is 5, and 41-5 = 36. The digital root of 36 is 3+6 = 9, which, as explained above, demonstrates that it is divisible by nine.
- The sum of the digits of 35967930 is 3+5+9+6+7+9+3+0 = 42, and 35967930-42 = 35967888. The digital root of 35967888 is 3+5+9+6+7+8+8+8 = 54, 5+4 = 9.
- Subtracting two base-10 positive integers that are transpositions of each other yields a number that is a whole multiple of nine. Examples:
- 41 - 14 = 27 (2 + 7 = 9)
- 36957930 - 35967930 = 990000, a multiple of nine.

This works regardless of the number of digits that are transposed. For example, the largest transposition of 35967930 is 99765330 (all digits in descending order) and its smallest transposition is 03356799 (all digits in ascending order); subtracting pairs of these numbers produces:

- 99765330 - 35967930 = 63797400; 6+3+7+9+7+4+0+0 = 36; 3+6 = 9.
- 99765330 - 03356799 = 96408531; 9+6+4+0+8+5+3+1 = 36; 3+6 = 9.
- 35967930 - 03356799 = 32611131; 3+2+6+1+1+1+3+1 = 18; 1+8 = 9.

Every prime in a Cunningham chain of the first kind with a length of 4 or greater is congruent to 9 mod 10 (the only exception being the chain 2, 5, 11, 23, 47).

Six recurring nines appear in the decimal places 762 through 767 of pi. This is known as the Feynman point.

If an odd perfect number is of the form 36k + 9, it has at least nine distinct prime factors.

If you divide a number by the amount of 9s corresponding to its number of digits, the number is turned into a repeating decimal.

Example: 274/999=0.274274274274

Nine is the binary complement of number six.

There are nine square feet in a square yard.

9, is the first odd number that is not prime.

Number 144 (1+4+4=9) is the square of 12, and is part of the Fibonacci numbers (see below). After 1 there are no other square numbers in the Fibonacci numbers other than 144!

The number 180 (1+8=9) is the maximum score in darts by scoring three treble 20s –

3 x (20x3) = 180 = 9.

Finding the Digital Root by Casting Nines
Casting out nines is a quick way of testing the calculations of sums, differences, products, and quotients of integers, known as long ago as the 12th Century.

You can cast out any number of digits that add up to 9. Therefore, to compute the digital root of 342718, cast out 3, 4, and 2 add up to 9 as do the digits 1 and 8. Cast these out and you will be left with 7 which is the digital root of this number.

$$342718$$
$$3+4+2 = 9 \quad 1+8 = 9$$
Digital root: 3+4+2+7+1+8 = 25. 2+5 = 7

If you subtract the numbers used in casting out nines from the original number, the result will be equal to digital root 7:

$$342718$$
$$\underline{-34218}$$
$$308500$$
$$3+8+5 = 16 \quad 1+6 = 7$$

Another example:

8624317
  8 + 1= 9, so cross out 8 and 1.
  6 + 3 = 9, so cross out 6 and 3.
  2 + 7 = 9, so cross out 2 and 7.
  Remaining number: **4**
  Therefore, the digital root is 4.
  8 + 6 + 2 + 4 +3 + 1 +7 = 31   3 +1 = **4!**

If you subtract the numbers used in casting out nines from the original number, the result will be equal to digital root 4:

$$8624317$$
$$-816327$$
$$7807990$$
$$7+8+7+9+9 = 40 = 4!$$

Subtracting the nines from a number results in a digital root equal to that of the original number.

Example: 48

48 - 9 = 39   39 - 9 = 30   30 - 9 = 21   21 - 9 = 12
12 - 9 = **3**

$$4+8 = 12$$
$$1+2 = 3!$$

Using the digital root to check addition
We can check an addition if it is right by finding the digital root of the numbers added, and comparing it to the digital root of the digital sum of the addition.

Example: 123 + 325 + 648
123 + 325 + 648 = 1096
DS: 1+0+9+6 = 16   DR: 1+6 = 7
DS 123: 1+2+3 = 6   DR = 6

DS 325: 3+2+5 = 10 DR = 1
DS 648: 6+4+8 = 18 DR = 9
Total DR: 6+1+9 = 16 = 7!

Number Pad (Numeric Keypad)
Look at your calculator or number-pad on the computer's keypad

$$\begin{array}{ccc} 7 & 8 & 9 \\ 4 & 5 & 6 \\ 1 & 2 & 3 \end{array}$$

Adding the numbers in threes, vertical or horizontal, top to bottom, bottom to top, left to right, right to left…

789	987	789	741	321	321
456	456	654	258	654	654
123	123	123	963	987	789
1368	1566	1566	1962	1962	1764

1368: 1+3+6+8 = 18 = 9
1566: 1+5+6+6 = 18 = 9
1962: 1+9+6+2 = 18 = 9
1764: 1+7+6+4 = 18 = 9

The following exercise reveals the digital root of nine; vertical, horizontal and diagonal:

$$\begin{array}{ccc} 7 & 8 & 9 \\ 4 & 5 & 6 \\ 1 & 2 & 3 \end{array}$$

789 x 741 = 731367 = 27 = 9
789 x 951 = 750339 = 27 = 9
789 x 963 = 950481 = 27 = 9
789 x 456 = 359784 = 36 = 9
789 x 123 = 97047 = 27 = 9
852 x 789 = 672228 = 27 = 9
852 x 456 = 388512 = 27 = 9

852 x 123 = 104796 = 27 = 9
852 x 951 = 810252 = 18 = 9
852 x 753 = 641556 = 27 = 9
Etc…

Square Root of 2
= 1.41421356237…
The DR of the first 9 numbers is equal to 9
1+4+1+4+2+1+3+5+6 = 27.   2+7 = 9

Magic Square of 11
110 124 136 148 160 172 184 196 208 220 232 = 1890
147 155 171 183 195 207 219 231 122 123 135 = 1890
182 194 204 218 230 121 133 134 146 158 170 = 1890
217 229 120 130 144 145 157 169 181 193 205 = 1890
131 143 155 156 166 180 192 204 216 228 119 = 1890
166 167 179 191 203 213 227 118 130 142 154 = 1890
190 202 214 226 117 129 139 153 165 177 178 = 1890
225 116 128 140 152 164 176 186 189 201 213 = 1890
139 151 163 175 187 199 200 212 222 115 127 = 1890
174 186 198 210 211 223 114 126 138 148 162 = 1890
209 221 222 113 125 137 149 1 61 173 185 195 = 1890
1890 1890 1890 1890 1890 1890 1890 1890 1890 1890
1890 =1+8+9 = 18.   1+8 = 9
1890 x 11 =20790   2+0+7+9+0 = 18   1+8 = 9
Diagonal: Top right, 232, to bottom left, 209
=1890 = 18 = 9!

Sudoku Nines

The Sudoku square has nine squares on each side.

5	3			7				
6			1	9	5			
	9	8					6	
8				6				3
4			8		3			1
7				2				6
	6					2	8	
			4	1	9			5
				8			7	9

In 2005, B. Felgenhauer, from Dresden in Germany, worked out that there are 6,670,903,752,021,072,936,960 different possibilities on a Sudoku grid...

6+6+7+9+3+7+5+2+2+1+7+2+9+3+6+9+6 = 90
9+0 = 9!

Geometry

Pi (π)
(3.14159265358979323846264338327950288419716939937510...)
Pi = 3.141592654
First 4 numbers: 3+1+4+1 = 9
Second 5 numbers: 5+9+2+6+5 = 27 = 9
First 9 numbers:
3+1+4+1+5+9+2+6+5 = 36 = 9
First 9 decimal numbers:
1+4+1+5+9+2+6+5+3 = 36 = 9

27 is the first 2-digit number that has a digital root of 9. It appears the first time after the 27th digit of the decimals. It is also surrounded by 3 and 9 (3x9=27)!

Circle

Degrees around the centre with various divisions:-
- 360 degrees. 3+6+0=9
- Half a circle: 180 degrees. 1+8+0=9
- Four quarters: 45 degrees. 4+5=9
- Sixteen sections: 22.5 degrees. 2+2+5 = 9
- Thirty-two sections: 11.25°. 1+1+2+5 = 9
- Sixty-four sections: 5.625. 5+6+2+5 = 18 = 9
- Hundred-twenty-eight sections: 2.8125
  2+8+1+2+5 = 18 = 9
- Two-hundred-fifty-six sections: 1.40625
  1+4+6+2+5 = 18 = 9

Polygons

The sum of all angles of any polygon will always equal 9.
Triangle: 3 angles = 180° = 1+8+0 = 9
Square: 4 angles = 90° x4 = 360 = 3+6+0 = 9

Polygon	Sides	Each Internal Angle	Sum of Internal Angles	Sum
Triangle	3	60	180	9
Square	4	90	360	9
Pentagon	5	108	540	9
Hexagon	6	120	720	9
Heptagon	7	128.6	900	9
Octagon	8	135	1080	9
Nonagon	9	140	1260	9
Decagon	10	144	1440	9
Undecagon	11	147.3	1620	9
Dodecagon	12	150	1800	9

The formula for **determining the sum of interior angles of a polygon** is n-2(180). n= number of sides. The sum of all angles of any polygon will always equal 9.

A 9-sided polygon is 9-2(180) or 7(180) or 1260. 1+2+6+0 = 9.

Time
- 1 year = 365 days + 6 hours. Number of hours per year = 365x24=8760 hours
  8760 + 6 = 8766 hours
  8+7+6+6 = 27. 2+7 = 9!
- Day = 24 hours, x 60 minutes = 1440 min = 9 = 86400 sec = 18 = 1+8 = 9
  Hour = 60 min = 60x60 = 3600 sec = 3+6 =9
- Minutes in 24 hours: 24 x 60 = 1440. 1+4+4= 9
- Seconds in an hour: 1x60x60 = 3600. 3+6 = 9 Seconds in a day: 24x60x60 = 86,400. 8+6+4 =18. 1+8 = 9
- Tetrasecond (Ts) = 1012 s =31689 years = 2 = 9

Geography

Arctic Circle: 66 33 N = 18 = 9

Antarctic Circle: 66 33 S = 18 = 9

Largest Desert: Sahara = 8397000 km2 = 27 = 9

Inclination of Earth is 23.4 degrees. 2+3+4= 9!

The longest river in South America is the Amazon, which is 6,300 km long (6+3=9).

The longest mountain range is the Andes in South America, which stretches for 4,500 miles (4+5=9).

Antarctica has a land mass of 9% of the globe.

The lowest point in Antarctica is 2,538 meters (2+5+3+8= 18 = 9).

The highest point in Africa is Mount Kilimanjaro, which is 5,985 meters (5+9+8+5= 27 = 9).

The deepest point in the oceans is the Mariana Trench in the Pacific Ocean, which is 36,198 feet deep (3+6+1+9+8= 27 = 9).

Archaeology

- The official length of the Great Wall of China is 7,200 kilometres. 7 + 2 = 9
- The base of the Great Pyramid of Khufu in Cairo, Egypt, measures 230.4 meters (2+3+4=9) which is 756 feet (7+5+6=18=9). The coordinates of the pyramid are: 29° 58' 45.3" N (2+9+5+8+4+5+3=36=9)! 31° 08' 03".69 E (3+1+8+3+6+9=30=3).
- The Pyramid of Khafra has a base of 215.28 meters (2+1+5+2+8=18=9) and a slope of 53° 10' (5+3+1=9).
- The Pyramid of Menkara has a coordinate of 29° 58' 21" N (2+9+5+8+2+1=27=9).

Science

Astronomy
- The sun is believed to be 4.5 billion years old (4+5=9).
- Before 2006 (when Pluto was officially designated as a non-planet), there were nine planets in the solar system.
- Messier object M9 is a magnitude 9.0 globular cluster in the constellation Ophiuchus.
- The New General Catalogue object NGC 9, a spiral galaxy in the constellation Pegasus
- The Saros number of the lunar eclipse series which began on -2501 June 26 and ended on -1149 September 16. The duration of Saros series 9 was 1352.2 years, and it contained 76 lunar eclipses.
- A full moon is nine times brighter than a half -moon.
- The earth is tilted at 23.4 degrees (2+3+4=9) and has a wobble due to its bulging centre, which gives us 4 seasons. Our earth's

north pole points through 12 constellations in a 25,920-year cycle (2+5+9+2= 18 = 9) broken up into 2160 years per constellation (2+1+6=9)
- Speed of light = 299700 km/sec
2+9+9+7+0+0 = 27 2+7 = 9
- First man in space: The first manned space flight lasted 1 Hour and 48 minutes, and was on April 12, 1961 by Yuri Gagarin, a Soviet cosmonaut.1 hour and 48 minutes is 108 minutes (1+8=9).
- The gravity constant on Earth is 9.81 meters/sec squared... 9+8+1=18= 9
- One Astronomical Unit (AU), the distance from Earth to the sun is 149,598,000 kilometres -
1+4+9+5+9+8 = 27 2+7 = 9
- Mass of the Earth: 5,400,000,000,000,000,000,000,000,000,000 g 5+4 = 9
- A light year, the distance light travels in vacuum in one year, is: 9,460,730,472,580.8 km
9+4+6+7+3+4+7+2+5+8+8 = 63 6+3 = 9
- The orbit of planet Mercury around the sun is 87.66 days. 8+7+6+6 = 27A light year, the distance light t2+7 = 9
- Jupiter has 63 moons. 6+3 = 9
- Uranus has 27 moons. 2+7 = 9
- Our solar system has a total of 162 moons. = 9
- From Pluto's perspective out at a distance of 33 times the typical Earth-Sun distance, the Sun would appear to be about 33 x 33 = 1,089 times fainter than it is from Earth. 1+8+9 = 18 = 9
- It takes light 4.5 hours to travel from Earth to Pluto. 4+5 = 9
- The New Horizons spacecraft left Earth faster than anything has before, and we've seen that it had to travel a tremendous distance to reach its destination. So how fast has it been traveling on its way to that destination? The answer is fast—really fast—about 9 miles per second or 32,400 miles per hour! Amazingly, this means that New Horizons is traveling towards Pluto much faster than Pluto is traveling around the Sun!
3+2+4 = 9

Chemistry

- The Nines scale is also sometimes used in describing the purity of bottled gases.
- Water expands by about 9% as it freezes.
- There are 90 (9+0 = 9) naturally occurring metals.
- Absolute zero is -273.15 °C. 2+7+3+1+5=18=9).
- Nine is the atomic number of fluorine. The atomic number is the number of protons in an element. Thus the atomic number is also known as the proton number.
- Chemical elements with increasing atomic numbers of nines:
  9- F- Fluorine   18- Ar- Argon   27-Co- Cobalt
  36-K-Krypton   45-Rh-Rhodium   54-Xe-Xenon
  63-Eu-Europium   72-Hf-Hafnium   81-Tl-Thallium   90-Th-Thorium   99-Es-Einsteinium
  108-Hs-Hassium   117-Uus-Ununseptium

Purity of Metals

Nines are an informal, yet common method of grading the purity of very fine precious metals such as platinum, gold and silver. Based on the system of millesimal fineness, a metal is said to be one nine or one nine fine if it is 900 fine, or 90% pure. A metal that is 990 fine is then described as two nines fine, and one that is 999 fine, is described as three nines fine. Thus, nines are a logarithmic scale of purity for very fine precious metals. The Nines scale is also sometimes used in describing the purity of bottled gases.

Physiology

- A human pregnancy normally lasts nine months, the basis of Naegele's rule.
- There are nine openings in the human body.
- The human adult heart beats 72 times (7+2=9) per minute.
- The lines in the palm of the left hand form the current Arabic numerals for 81 (8+1=9), and the right palm forms 18!
- The first cry of a new born baby averages 432Hz (4+3+2=9).

Computing
- The lowest recommended refresh rate for a computer screen is 72Hz (7+2=9).
- The standard resolution of a computer screen is 72 dpi (dots per inch).

Sports

Baseball
- In baseball, nine represents the right fielder's position.
- The number of innings in a regulation, non-tied game of baseball.
- The number of players on the field including the pitcher.
- The number worn by Roy Hobbs in the movie The Natural.
- The score of a forfeit in a league where the game would ordinarily last nine innings is 9-0.
- NINE: A Journal of Baseball History and Culture published by the University of Nebraska Press.
- A baseball team is made up of nine players.

Billiards
Nine-ball is the standard professional pocket billiards variant played in the United States.

Rugby
- In rugby league, the jersey number assigned to the hooker.
- In rugby union, the number worn by the starting scrum-half.

Football (Soccer)
In association football (soccer), the center-forward/striker traditionally (since at least the fifties) wears the number 9 shirt.

Golf
The most common par for an 18-hole (1+8=9) golf course is 72 (7+2=9).

Other Games

The game of skittles or ninepin bowling is many hundreds of years old. The pins are set up in a diamond formation and players throw the ball (or 'cheese') at them. In the nineteenth century some American states passed laws banning the game because bets were often placed on it. But these laws were evaded by adding a tenth pin. As a result tenpin bowling is now the far more popular game.

Languages

The longest one-syllable word in the English language is 9-letter long: "screeched".

In French the word 'neuf' means both nine and new. In German, the words for nine and new are 'neun' and 'neu', and in Spanish, 'nueve' and 'nuevo'.

The Cambodian alphabet is the longest in the world, made of 72 letters.

Redivider, with nine letters, is the longest palindromic word in the English language. A palindromic word has the same sequence of letters backwards or forwards.

Nine in different languages:

The list is very long. Please refer to the following website: https://en.wikipedia.org/wiki/List_of_numbers_in_various_languages

Music

Beethoven wrote 9 symphonies, after which he died. To this day, a superstition among many musical composers forbids the numbering of a symphony past the number 9. Mahler wrote more symphonies, but never named any one of them, number 9.

Symphony No.9
1. Mozart, Wolfgang Amadeus. 1772.
2. Beethoven, Ludwig van. 1824.

3. Dvorak, Antonin. 1893.
4. Mahler, Gustav. 1908-1909.
5. Maslanka, David. 2011.
6. Bruckner, Anton. 1896.
7. Schubert, Franz. 1840.
8. Shostakovich, Dimitri. 1945.
9. Haydn, Michael. 1766.
10. Glass, Philip. 2010-2011.
11. Davies, Peter Maxwell. 2011-2012.
12. Henze, Hans Werner. 1997.
13. Glazunov, Alexander. 1910.
14. Arnold, Malcolm. 1986.
15. Schnittke, Alfred. 1996.
16. Myaskovsky, Nikolai. 1926-1927.
17. Pettersson, Allen. 1979.
18. Milhaud, Darius. 1959.
19. Williams, Vaughan. 1956-1957.
20. Simpson, Robert. 1985-1987.

Telecommunication

International telephone codes:

1. Argentina 54 (=9)
2. Aruba 297 (=18=9)
3. Bosnia & Herzegovina 387 (=18=9)
4. Cambodia 855 (=18=9)
5. Central Democratic Republic 243 (=9)
6. Cote D'Ivoire 225 (=9)
7. Denmark 45 (=9)
8. French Guiana 594 (=18=9)
9. Honduras 504 (=9)
10. Hungary 36 (=9)
11. Israel 972 (=18=9)
12. Japan 81 (=9)
13. Lichtenstein 423 (=9)

14. Madagascar 261 (=9)
15. Nigeria 234 (=9)
16. Papua New Guinea 675 (=18=9)
17. Philippines 63 (=9)
18. Portugal 351 (=9)
19. San Marino 378 (=18=9)
20. Somalia 252 (=9)
21. South Sudan 27 (=9)
22. Syria 963 (=18=9)
23. Tunisia 216 (=9)
24. Turkey 90 (=9)

Miscellaneous

- Nine judges sit on the United States Supreme Court.
- Nine is the triad of triads, hence the first odd square number...
- The ratio between 9 & 8 defines the crucial whole tone in music from which the scale emerges...
- A standard work day begins at 9 am
- The standard dimensions of a play card are: 6.35 x **9** cm
- 'Cats have nine lives' is an expression of the cat's ability to avoid and escape danger.
- The 'cat of nine tails' was a whip used to force discipline aboard ships. It had nine knotted strands which cut into the skin and flesh.
- The first man to fly a plane through the sound barrier was Chuck Yeager, an American pilot, in a Bell XS-1 on 14 October 1947, at an altitude of 45,000 feet.
  14 Oct. 1947: 1+4+1+0+1+9+4+7 = 27   2+7 = 9
  45,000: 4+5 = 9!
- The Millau Viaduct, in the Massif Central of France, was the world's tallest road bridge when it opened on 14 December 2004. Its height is 270 meters (2+7=9) which is equal to 891 feet (8+9+1=18=9)... The bridge is suspended on seven pillars, the tallest of which is 342m high (3+4+2=9)...

- Number 360 (3+6=9) is the smallest number with 24 factors, making it very possible to divide up into months, days, minutes, seconds, etc…
- The Colt 45 (4+5=9) is one of the most famous revolvers ever made. It was designed for the US Cavalry in the 19th century and assumed the nickname of the 'Peacemaker'.
- Number 216 (2+1+6=9) is the smallest cube that is the sum of 3 cubes:
  3³ + 4³ + 5³ = 27 + 64 + 125 = 216 = 9.
- The Rubik's cube has 54 (5+4=9) coloured squares – six faces of nine squares.
- There are 36 black keys on a piano.
- There are 108 windows in the dome of the US Capitol Building.
- The **Boeing 747** is a wide-body commercial jet airliner and cargo aircraft, often referred to by its original nickname, *Jumbo Jet*, or *Queen of the Skies*. Its distinctive "hump" upper deck along the forward part of the aircraft makes it among the world's most recognizable aircraft, and it was the first wide-body produced. First flown commercially in 1970, the 747 held the passenger capacity record for 37 years.
- The Vatican City sits on 108 acres.
- The International Standard Book Number (ISBN) was first generated in 1967 based upon the 9-digit SBN created in 1966.
- Again, 216 is the constant in a 3x3 square. Each row and column multiplies to 216!

2	9	12
36	6	1
3	4	18

Horizontal:
2x9x12 = 216   36x6x1 = 216   3x4x18 = 216

Vertical:
2x36x3 = 216   9x6x4 = 216   12x1x18 = 216

Diagonal:
2x6x18 = 216   12x6x3 = 216
(2+1+6 = 9)

- If you multiply nine by any whole number (except zero), and repeatedly add the digits of the answer until it's just one digit, digital root, you will end up with nine:

  2 × 9 = 18 (1 + 8 = 9)
  3 × 9 = 27 (2 + 7 = 9)
  9 × 9 = 81 (8 + 1 = 9)
  121 × 9 = 1089 (1 + 0 + 8 + 9 = 18; 1 + 8 = 9)
  234 × 9 = 2106 (2 + 1 + 0 + 6 = 9)
  578329 × 9 = 5204961 (5 + 2 + 0 + 4 + 9 + 6 + 1 = 27 (2 + 7 = 9))
  482729235601 × 9 = 4344563120409 (4 + 3 + 4 + 4 + 5 + 6 + 3 + 1 + 2 + 0 + 4 + 0 + 9 = 45 (4 + 5 = 9))
  (Exception) 0 x 9 = 0 (0 is not equal to 9)

References

1- Guedj, Denis. Numbers the Universal Language.London.1998. ISBN 0-500-30080-1
2- Schimmel, Annemarie. Mystery of Numbers. Oxford. 1993. ISBN 0-19-506303-1
3- Al-Khalili, Jim. Pathfinders. London. 2010. ISBN 978-0-141-03836-0
4- Ifrah, George. The Universal History of Numbers. London.1994. ISBN 1-86046-324-x
5- Kline, Morris. Mathematical Thought from Ancient to Modern Times, Oxford University Press, 1972.
6- Dantzig, Tobias. Number, the language of science: a critical survey written for the cultured non-mathematician, New York, The Macmillan Company, 1930.
7- Flegg, Graham. Numbers-Their History and Meaning. Barnes & Noble Books, 1993
8- Wells, David. The Penguin Dictionary of Curious and Interesting Numbers. Penguin, 1987.
9- Wells, D. The Penguin Book of Curious and Interesting Mathematics by, Penguin Books, 1997.
10- Cullerne, John. Mathematics Basic Facts, Harper Collins Publishers, 1998
11- Gullberg, Jan. Mathematics from the Birth of Numbers. W.W. Norton & Co., 1996
12- Debaene, Stanislas. The Number Sense. Oxford University Press, New York, 1997
13- Balmond, Cecil. "Number 9, the search for

the sigma code" 1998, Prestel 2008.
ISBN 37913-1933-7
ISBN 978-3-7913-1933-9.
14- Devi, Shakuntala. The Book of Numbers. India. 2008. ISBN 978-81-222-00006-5
15- Law, Steven. A Brief History of Numbers and Counting, Part I & 2. Yale Centre for the Study of Globalisation. 2012
16- Boyer, C.B. Fundamental Steps in the Development of Numeration.
17- Conant, Levi Leonard. The Number Concept: Its Origin and Development. New York.1931. (EBook - 16449. 2005).
18- Glynne-Jones, Tim. The Book of Numbers. London. 2007. ISBN 978-1-84837-190-3
19- Rees, Martin. Just Six Numbers. London, 999. ISBN 0 75381 022 0
20- Crilly, Tony. 50 Mathematical Ideas You Really Need to Know. London. 2007. ISBN: 978-1-84724-008-8
21- Koshy, Thomas. Fibonacci and Lucas Numbers with Applications.
ISBN: 978-0-471-39969-8
22- Olsen, Scott. The Golden Section. Glastonbury. 2006. ISBN: 1 904263 47X
23- Pickover, Clifford A. Wonders of Numbers. Oxford. 2001. ISBN 0-19-515799-0
24- **E-books**
  24.01- Mathematical Essays and Recreations. Hermann Schubert
  Translator: Thomas J. McCormack
  Release Date: May 9, 2008
  [EBook #25387]
  Original Book: Chicago, 1898
  24.02- Essays on the Theory of Numbers
  Author: Richard Dedekind

Translator: Wooster Woodruff Beman
Release Date: April 8, 2007
[EBook #21016]
Original Book: Chicago, 1901

24.03- An Introduction to Mathematics
Author: Alfred North Whitehead
Release Date: December 6, 2012
[EBook #41568]

24.04 - The Hindu-Arabic Numerals
Author: David Eugene Smith &Louis Charles Karpinski.
Release Date: September 14, 2007
[EBook#22599]

24.05 -The Theory of Numbers
Author: Robert D. Carmichael
Release Date: April 8, 2013
[EBook #13693]
Original Book: New York, 1914

24.06 –The Number Concept, Its Origin and Development
Author: Levi Leonard Conant
Release Date: August 5, 2005
[EBook #16449]
Original Book: New York, 1931

24.07 - Philosophy and Fun of Algebra
Author: Mary Everest Boole
Release Date: September 12, 2004
[EBook #13447]
Date last updated December 3, 2005

24.08 - Mathematical Recreations and Essays.
Author: W. W. Rouse Ball
Release Date: October 8, 2008
[EBook #26839]
Original Book: Oxford, 1892

24.09 - Mathematical Essays and

Recreations
Author: Hermann Schubert
Translator: Thomas J. McCormack
Release Date: May 9, 2008
[EBook #25387]
Original Book: Chicago, 1898

24.10- Essays on the Theory of Numbers
Author: Richard Dedekind
Translator: Wooster Woodruff Beman
Release Date: April 8, 2007
[EBook #21016]
Original Book: Chicago, 1901

24.11 - An Introduction to Mathematics
Author: Alfred North Whitehead
Release Date: December 6, 2012
[EBook #41568]

24.12 - Introduction to Mathematical Philosophy
Author: Bertrand Russell
Release Date: February 17, 2013
[EBook #41654]
Original Book: London, 1919

25- Wikipedia, the free encyclopaedia.
(http://en.wikipedia.org/)

26- Project Gutenberg.
(http://www.gutenberg.org/)

www.ingramcontent.com/pod-product-compliance
Lightning Source LLC
Chambersburg PA
CBHW031050180526
45163CB00002BA/765